Analysis and Control Using the
Lambert W Function

TIME-DELAY
SYSTEMS

T0350180

Analysis and Control Using the
Lambert W Function

TIME-DELAY SYSTEMS

Sun Yi
Patrick W Nelson
A Galip Ulsoy

University of Michigan, Ann Arbor, USA

World Scientific

NEW JERSEY · LONDON · SINGAPORE · BEIJING · SHANGHAI · HONG KONG · TAIPEI · CHENNAI

Published by

World Scientific Publishing Co. Pte. Ltd.

5 Toh Tuck Link, Singapore 596224

USA office: 27 Warren Street, Suite 401-402, Hackensack, NJ 07601

UK office: 57 Shelton Street, Covent Garden, London WC2H 9HE

British Library Cataloguing-in-Publication Data
A catalogue record for this book is available from the British Library.

TIME-DELAY SYSTEMS
Analysis and Control Using the Lambert W Function

ISBN-13 978-981-4307-39-0
ISBN-10 981-4307-39-4

Printed in Singapore.

To our families, with love

Preface

This book is a collection of recent research work on the development of an analytical approach for solutions of delay differential equations via the Lambert W function. It, also, includes methods for analysis and control based on the solutions, and their applications to mechanical and biological systems.

Delay differential equations represent systems that include inherent time-delays in the system or a deliberate introduction of time-delays for control purposes. Such time-delays, common in systems in engineering and science, can cause some significant problems (e.g., instability and inaccuracy) and, thus, limit and degrade the achievable performance of controlled systems. However, due to their innate complexity including infinite-dimensionality, it is not feasible to analyze such systems with classical methods developed for ordinary differential equations (ODEs).

The research presented in this book uses the Lambert W function to obtain free and forced closed-form solutions to such systems. Hence, it provides a more analytical and effective way to treat time-delay systems. The advantage of this approach stems from the fact that the closed-form solution is an infinite series expressed in terms of the parameters of the system. Thus, one can explicitly determine how the parameters are involved in the solution. Furthermore, one can determine how each system parameter affects the eigenvalues of the system. Also, each eigenvalue in the infinite eigenspectrum is associated individually with a branch of the Lambert W function.

The Lambert W function-based approach for the analytical solution to systems of delay differential equations (DDEs) had previously been developed for homogeneous first-order scalar and some special cases of systems of delay differential equations using the Lambert W function as introduced in Chapter 1. In Chapter 2, the analytical solution is extended to the more general case where the coefficient matrices do not necessarily commute, and to the nonhomogeneous case. The solution is in the form of an infinite series of modes written in terms of the matrix Lambert W function. The derived solution is used to investigate the stability of time-delay systems via dominant eigenvalues in terms of the Lambert W function. It is also applied to the regenerative machine tool chatter problem of a manufacturing process in Chapter 3. Based on the solution form in terms of the matrix Lambert W function, algebraic conditions and Gramians for controllability and observability of DDEs are derived in a manner analogous to the well-known controllability and observability results for ODEs in Chapter 4. In Chapter 5, the problem of feedback controller design via eigenvalue assignment for linear time-invariant time-delay systems is considered. The method for eigenvalue assignment is extended to design robust controllers for time-delay systems with uncertainty and to improve transient response in Chapter 6. For systems where all state variables cannot be measured directly, a new approach for observer-based feedback control is developed and applied to diesel engine control in Chapter 7. In Chapter 8, the approach using the Lambert W function is applied to analyze a HIV pathogenesis dynamic model with an intracellular delay.

The authors hope that this book will be of interest to graduate students and researchers in engineering and mathematics who have special interest in studying the properties, and in designing controllers, for time-delay systems.

The authors are pleased to acknowledge support for this research by a research grant (# 0555765) from the National Science Foundation.

S. Yi
P. W. Nelson
A. G. Ulsoy

Contents

Chapter 1

Introduction

1.1 Motivation

Time-delay systems (TDS) arise from inherent time-delays in the components of the systems, or from the deliberate introduction of time-delays into the systems for control purposes. Such time-delays occur often in systems in engineering, biology, chemistry, physics, and ecology (Niculescu, 2001). Time-delay systems can be represented by delay differential equations (DDEs), which belong to the class of functional differential equations, and have been extensively studied over the past decades (Richard, 2003). Such time-delays can limit and degrade the achievable performance of controlled systems, and even induce instability. Time-delay terms lead to an infinite number of roots of the characteristic equation, making systems difficult to analyze with classical methods, especially, in checking stability and designing stabilizing controllers. Thus, such problems are often solved indirectly by using approximations. A widely used approximation method is the Padé approximation, which is a rational approximation and results in a shortened fraction as a substitute for the exponential time-delay term in the characteristic equation. However, such an approach constitutes a limitation in accuracy, can lead to instability of the actual system and induce non-minimum phase and, thus, high-gain problems (Silva and Datta, 2001). Prediction-based methods (e.g., Smith predictor (Smith, 1957), finite spectrum assignment (FSA) (Zhong, 2006), and adaptive Posicast (Niculescu and Annaswamy, 2003)) have been used to stabilize time-delay systems by transforming the problem into a non-delay system. Such methods require model-based calculations, which may cause unexpected errors when applied to a real system. Furthermore, safe implementation of such methods is still an open problem due to computational issues. Controllers have also been designed using the Lyapunov framework (e.g., linear matrix inequalities

(LMIs) or algebraic Riccati equations (AREs)) (Gu and Niculescu, 2006; Liu, 2003). These methods require complex formulations, and can lead to conservative results and possibly redundant control.

To find more effective methods, an analytic approach to obtain the complete solution of systems of delay differential equations based on the concept of the Lambert W function, which has been known to be useful to analyze DDEs (Corless *et al.*, 1996), was developed in (Asl and Ulsoy, 2003). The solution has an analytical form expressed in terms of the parameters of the DDE and, thus, one can explicitly determine how the parameters are involved in the solution and, furthermore, how each parameter affects each eigenvalue and the solution. Also, each eigenvalue is associated individually with a particular 'branch' of the Lambert W function. In this book, the analytical approach using the Lambert W function is extended to general systems of DDEs and non-homogeneous DDEs, and compared with the results obtained by numerical integration. The advantage of this approach lies in the fact that the form of the solution obtained is analogous to the general solution form of ordinary differential equations, and the concept of the state transition matrix in ODEs can be generalized to DDEs using the concept of the matrix Lambert W function. This suggests that some approaches for analysis and control used for systems of ODEs, based on concept of the state transition matrix, can potentially be extended to systems of DDEs. These include analysis of stability, controllability and observability, and methods for eigenvalue assignment for linear feedback controller design with an observer, and extension to robust stability and time-domain specifications. Also, the approaches developed based on the proposed solution method are applied to time-delay systems in engineering and biology as discussed in subsequent chapters.

1.2   Background

1.2.1   *Delay differential equation*

Delay differential equations are also known as difference-differential equations, were initially introduced in the 18th century by Laplace and Condorcet (Gorecki *et al.*, 1989). Delay differential equations are a type of differential equation where the time derivatives at the current time depend on the solution, and possibly its derivatives, at previous times. A class of such equations, that involve derivatives with delays as well as the solution itself have historically been called *neutral* DDEs (Hale and Lunel,

1993). In this book only *retarded* DDEs, where there is no time-delay in the derivative terms, are considered.

The basic theory concerning stability and works on fundamental theory, e.g., existence and uniqueness of solutions, was presented in (Bellman and Cooke, 1963). Since then, DDEs have been extensively studied in recent decades and a great number of monographs have been published including significant works on dynamics of DDEs by Hale and Lunel (1993), on stability by Niculescu (2001), and so on. The reader is referred to the detailed review in (Richard, 2003; Gorecki *et al.*, 1989; Hale and Lunel, 1993). The interest in the study of DDEs is caused by the fact that many processes have time-delays and have been modeled for better fidelity by systems of DDEs in the sciences, engineering, economics, etc. (Niculescu, 2001). Such systems, however, are still not feasible to precisely analyze and control, thus, the study of systems of DDEs has actively been conducted during recent decades (Richard, 2003).

1.2.2 *Lambert W function*

Introduced in the 1700s by Lambert and Euler (Corless *et al.*, 1996), the Lambert W function is defined to be any function, $W(H)$, that satisfies

$$W(H)e^{W(H)} = H \qquad (1.1)$$

The Lambert W function is complex valued, with a complex argument, H, and has an infinite number of branches, W_k, where $k = -\infty, \cdots, -1, 0, 1, \cdots, \infty$ (Asl and Ulsoy, 2003). Figure 1.1 shows the range of each branch of the Lambert W function. For example, the real part of the principal branch, W_0, has a minimum value, -1. The principal and all other branches of the Lambert W function in Eq. (1.1) can be calculated analytically using a series expansion (Corless *et al.*, 1996), or alternatively, using commands already embedded in the various commercial software packages, such as Matlab, Maple, and Mathematica.

An analytic approach to obtain the complete solution of systems of delay differential equations based on the concept of the Lambert W function was developed by Asl and Ulsoy (2003). Consider a first-order scalar homogenous DDE:

$$\begin{aligned} \dot{x}(t) &= ax(t) + a_d x(t-h), \, t > 0 \\ x(0) &= x_0, & t = 0 \\ x(t) &= g(t), & t \in [-h, 0) \end{aligned} \qquad (1.2)$$

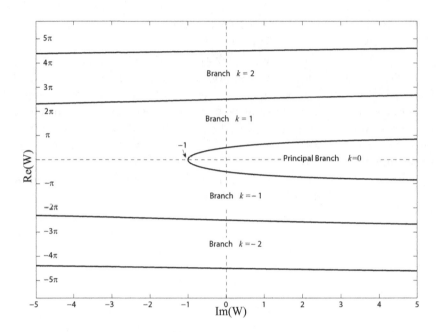

Fig. 1.1   Ranges of each branch of the Lambert W function (Corless *et al.*, 1996). Note that real part of the principal branch, W_0, is equal to or larger than -1.

Instead of a simple initial condition as in ODEs, two initial conditions need to be specified for DDEs: a preshape function, $g(t)$, for $-h \leq t < 0$ and initial point, x_0, at time, $t = 0$. This permits a discontinuity at $t = 0$, when $x_0 \neq g(t = 0)$. The quantity, h, denotes the time-delay. The solution to Eq. (1.2) can be derived in terms of an infinite number of branches of the Lambert W function, defined in Eq. (1.1), (Asl and Ulsoy, 2003):

$$x(t) = \sum_{k=-\infty}^{\infty} e^{S_k t} C_k^I, \quad \text{where } S_k = \frac{1}{h} W_k(a_d h e^{-ah}) + a \qquad (1.3)$$

The coefficient, C_k^I, is determined numerically from the preshape function, $g(t)$, and initial state, x_0, defined in the Banach space as described by Asl and Ulsoy (2003). The analytic methods to find the coefficient, C_k^I and the numerical and analytic methods for other coefficients for non-homogeneous and higher order of DDEs are also developed in a subsequent chapter. Note that, unlike results by other existing methods, the solution in Eq. (1.3) has an analytical form expressed in terms of the parameters of the DDE in Eq. (1.2), i.e., a, a_d and h. One can explicitly determine how the parameters are involved in the solution and, furthermore, how each parameter affects each eigenvalue and the solution. Also, each eigenvalue is distinguished by k, which indicates the branch of the Lambert W function as seen in Eq. (1.3).

1.3 Scope of This Document

This book presents the derivation of solutions of systems of DDEs, and the development of methods to analyze and control time-delay systems with application to systems in engineering and biology. This new technique allows one to study how the parameters in time-delay systems are involved in the solution, which is essential to investigate system properties, such as stability, controllability, observability, and sensitivity. Finally, controllers for time-delay systems, with observers, are designed via eigenvalue assignment to improve robust stability and to meet time-domain specifications as well as to stabilize unstable systems (See Fig. 1.2).

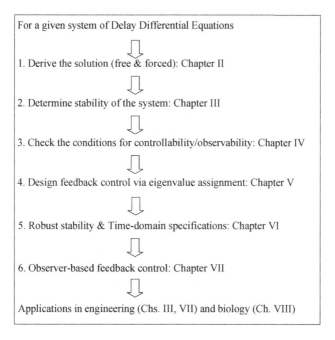

Fig. 1.2   The matrix Lambert W function-based approach: using the approach developed in this research, the steps in the figure, which are standard for systems of ODEs, become tractable for DDEs.

Because each chapter of this book is based on manuscripts that have been published in a journal or conference, the background material for each is included in the relevant chapters. The remaining chapters are summarized as follows.

Chapter II: "Solutions of Systems of DDEs via the Matrix Lambert W Function", which was published in the *Dynamics of Continuous, Discrete and Impulsive Systems (Series A)* (Yi *et al.*, 2007d) and an early version of this work was presented in part at the 2006 American Control Conference (Yi and Ulsoy, 2006) and in part at the 2006 IEEE Conference on Decision and Control (Yi *et al.*, 2006b). Previously, an approach for the analytical solution to systems of DDEs had been developed for homogeneous scalar and some special cases of systems of delay differential equations using the Lambert W function (Asl and Ulsoy, 2003). In this chapter, the approach is extended to the more general case where the coefficient matrices in a system of DDEs do not necessarily commute, and to the nonhomogeneous cases. The solution is in the form of an infinite series of modes written in terms of the *matrix* Lambert W function. The form of the obtained solution has similarity to the concept of the state transition matrix in linear ordinary differential equations, enabling its use for general classes of linear delay differential equations. Examples are presented to illustrate the new approach by comparison to numerical methods. The analytical solution in terms of the Lambert W function is also presented in the Laplace domain to reinforce the analogy to ODEs.

Chapter III: "Stability of Systems of Delay Differential Equations via the Matrix Lambert W Function: Application to machine tool chatter," which was published in the *Mathematical Biosciences and Engineering* (Yi *et al.*, 2007b) and an earlier version of this work was presented at the 2006 ASME International Conference on Manufacturing Science and Engineering (Yi *et al.*, 2006a). This chapter investigates stability of systems of DDEs using the solution derived in terms of the parameters of systems in Chapter II. By applying the matrix Lambert W function-based approach to the chatter equation, one can solve systems of DDEs in the time domain, obtain dominant eigenvalues, and check the stability of the system. With this method one can obtain ranges of preferred operating spindle speed that do not cause chatter to enhance productivity of processes and quality of products. The new approach shows excellent accuracy and certain other advantages, when compared to existing graphical, computational and approximate methods.

Chapter IV: "Controllability and Observability of Systems of Linear Delay Differential Equations via the Matrix Lambert W Function," which was published in the *IEEE Transactions on Automatic Control* (Yi *et al.*, 2008a) and an earlier version of this work was presented at the 2007 American Control Conference (Yi *et al.*, 2007a). Controllability and observability of linear time-delay systems has been studied, and various definitions and criteria have been presented since the 1960s (Malek-Zavarei and Jamshidi, 1987), (Yi *et al.*, 2008a). However, the lack of an analytical solution approach has limited the applicability of the existing theory. In this chapter, based on the solution form in terms of the matrix Lambert W function, algebraic conditions and Gramians for controllability and observability of DDEs were derived in a manner analogous to the well-known controllability and observability results for the ODE case. The controllability and observability Gramians indicate how controllable and observable the corresponding states are, while algebraic conditions tell only whether a system is controllable/observable or not. With the Gramian concepts, one can determine how the changes in some specific parameters of the system affect the controllability and observability of the system via the resulting changes in the Gramians. Furthermore, for systems of ODEs, a balanced realization in which the controllability Gramian and observability Gramian of a system are equal and diagonal was introduced in (Moore, 1981). Using the Gramians defined in this chapter, the concept of the balanced realization has been extended to systems of DDEs for the first time.

Chapter V: "Eigenvalue Assignment via the Lambert W Function for Control for Time-Delay Systems," which is in press in the *Journal of Vibration and Control* (Yi *et al.*, 2010b) and an earlier version of this work was presented at the 2007 ASME International Design Engineering Technical Conferences (Yi *et al.*, 2007c). In this chapter, the problem of feedback controller design via eigenvalue assignment for linear time-invariant systems of linear delay differential equations with a single delay is considered. Unlike ordinary differential equations, DDEs have an infinite eigenspectrum and it is not feasible to assign all closed-loop eigenvalues. However, one can assign a critical subset of them using a solution to linear systems of DDEs in terms of the matrix Lambert W function. The solution has an analytical form expressed in terms of the parameters of the DDE, and is similar to the state transition matrix in linear ODEs. Hence, one can extend controller design methods developed based upon the solution form of systems of ODEs to systems of DDEs, including the design of feedback controllers via eigenvalue assignment. Such an approach is presented here, illustrated using some examples, and compared with other existing methods.

Chapter VI: "Robust Control and Time-Domain Specifications," which was published in the *Journal of Dynamic Systems, Measurement, and Control* (Yi *et al.*, 2010c) and an earlier version was presented at the 2008 American Control Conference (Yi *et al.*, 2008c). One of the main concerns in designing controllers is to maintain robust stability against uncertainty in the models. When uncertainty exists in the coefficients of the system, a robust control law, which can guarantee stability, is required. To realize robust stabilization, after calculating the allowable size of uncertainty (i.e., norms of the uncertainty matrices), the rightmost eigenvalues are placed at an appropriate distance from the imaginary axis to maintain stability even with the uncertainty in the coefficients. An algorithm for eigenvalue assignment for systems of DDEs, based upon the Lambert W function, is devised for the problem of robust control design for perturbed systems of DDEs. With this algorithm, after considering the size of the allowable uncertainty in the coefficients of systems of DDEs via the stability radius analysis and comparing the stability radius and real uncertainty in parameters, the appropriate positions of the rightmost eigenvalues for robust stability are chosen such that the stability radius of the controlled system is larger than the size of uncertainty. Corresponding to the calculated positions, one can find appropriate gains of the linear feedback controller by assigning the rightmost eigenvalues using the method introduced in Chapter V. By moving the rightmost eigenvalues, the stability radius is increased to be larger than the size of uncertainty. The procedure presented in this chapter can be applied to uncertain systems, where uncertainty in the system parameters cannot be ignored. Also, the method developed in Chapter V makes it possible to assign simultaneously the real and imaginary parts of a critical subset of the eigenspectrum for the first time. Therefore, similar guidelines to those for systems of ODEs to improve transient response and to meet time-domain specifications, can be developed and used for systems of DDEs via eigenvalue assignment.

Chapter VII: "Design of Observer-Based Feedback Control for Time-Delay Systems with Application to Automotive Powertrain Control," was published in the *Journal of the Franklin Institute* (Yi *et al.*, 2010a), and the Proceedings of 2009 ASME Dynamic Systems and Control Conference (Yi *et al.*, 2009). In this chapter, a new approach for observer-based feedback control of time-delay systems is developed. The approach, based on the Lambert W function, is used to control time-delay systems by designing an observer-based state feedback controller via eigenvalue assignment. The designed observer provides estimation of the state, which converges asymptotically to the actual state, and is then used for state feedback control. It is shown that the separation principle applies as for the case of ODE's. The feedback controller and the observer take simple linear forms and, thus, are easy to implement when compared to nonlinear methods. This new approach is applied, for illustration, to the control of a diesel engine to achieve improvement in fuel efficiency and reduction in emissions. The simulation results show excellent closed-loop performance.

Chapter VIII: "Eigenvalues and Sensitivity Analysis for a Model of HIV Pathogenesis with an Intracellular Delay", which is based upon a manuscript presented at the 2008 ASME Dynamic Systems and Control Conference (Yi *et al.*, 2008b). During the past decade significant research has been aimed at better understanding of the human immunodeficiency virus (HIV), and the use of mathematical modeling to interpret experimental results has made a significant contribution. However, time-delays, which play a critical role in various biological models, are still not amenable to many traditional analysis methods. In this chapter, the approach using the Lambert W function is applied to handle the time-delay in a HIV pathogenesis dynamic model. Dominant eigenvalues in the infinite eigenspectrum of these time-delay systems are obtained and used to understand the effects of the parameters of the model on the immune system. Also, the result is extended to analyze the sensitivity of the eigenvalues with respect to the parameters in the HIV model. The research makes it possible to know which parameters are more influential than others, and the information obtained is used to investigate the HIV dynamic system analytically.

1.4 Original Contributions

The original contributions of the research documented in this book for time-delay systems can be summarized as follows:

(1) Derivation of free and forced solutions of general systems of DDEs, which take an analytical form in terms of systems parameters and, thus, enable understanding of how they are involved in the solutions and dynamics. (Chapter II)

(2) Determination of stability of infinite dimensional systems of DDEs based upon a finite, but dominant, number of eigenvalues. (Chapter III)

(3) Development of conditions for controllability and observability, which indicate how controllable and observable for a system is, for DDEs. The conditions also indicate how the change in some specific parameters of the system affect the controllability and observability of the systems and, furthermore, can be used for balancing a realization. (Chapter IV)

(4) A method for the design of feedback controllers via eigenvalue assignment to assign dominant eigenvalues to desired positions. (Chapter V)

(5) Algorithms for robust stabilization and achievement of time-domain specifications. (Chapter VI)

(6) Design of observer-based feedback controllers for time-delay systems where all state variables cannot be measured directly. (Chapter VII)

(7) Dominant eigenvalues in the infinite eigenspectrum of these time-delay systems are obtained and used to understand the effects of the parameters of a HIV pathogenesis dynamic model. (Chapter VIII)

Chapter 2

Solutions of Systems of DDEs via the Matrix Lambert W Function

An approach for the analytical solution to systems of delay differential equations (DDEs) has been developed for homogeneous scalar and some special cases of systems of DDEs using the Lambert W function. In this chapter, such an approach is extended to the more general case where the coefficient matrices in a system of DDEs do not commute, and to the nonhomogeneous case. The solution is in the form of an infinite series of modes written in terms of the matrix Lambert W function. The form of the obtained solution has similarity to the concept of the state transition matrix in linear ordinary differential equations (ODEs), enabling its use for general classes of linear delay differential equations. Examples are presented to illustrate the new approach by comparison to numerical methods. The analytical solution in terms of the Lambert W function is also presented in the Laplace domain to reinforce the analogy to ODEs.

2.1 Introduction

Time-delay systems are systems in which a significant time-delay exists between the applications of input to the system and their resulting effect. Such systems arise from an inherent time delay in the components of the system or from a deliberate introduction of time delay into the system for control purposes. Such time-delay systems can be represented by delay differential equations, which belong to the class of functional differential equations, and have been extensively studied over the past decades (Richard, 2003). The principal difficulty in studying DDEs results from their special transcendental character. Delay problems always lead to an infinite spectrum of frequencies. The determination of this spectrum requires a corresponding determination of roots of the infinite-dimensional characteristic equation, which is not feasible, in general, by using standard methods developed for systems of ODEs. For this reason, instead of obtaining closed-form solutions, systems of DDEs are often handled using numerical methods, asymptotic solutions, and graphical approaches mainly for stability analysis and design of controllers. For a more detailed discussion and comparison of such existing methods, the reader is referred to (Richard, 2003; Asl and Ulsoy, 2003; Gorecki *et al.*, 1989; Yi *et al.*, 2010b) and the references therein.

During recent decades, the spectral decomposition methods for solutions of DDEs in terms of generalized eigenfunctions have been developed (Banks and Manitius, 1975; Bellman and Cooke, 1963; Bhat and Koivo, 1976a,b; Hale and Lunel, 1993), and applied to control problems. Recently, based on the concept of the Lambert W function, an analytic approach to obtain the solution of homogeneous scalar delay differential equations has been developed by Asl and Ulsoy (2003) and Corless *et al.* (1996). That is, as introduced in Section 1.2.2, for the first-order scalar homogenous DDE in Eq. (1.2), the solution in Eq. (1.3) is derived in terms of an infinite number of branches of the Lambert W function, W_k. Note that, unlike results by other existing methods (see, e.g., (Richard, 2003) and the references therein), the solution in (1.3) has an analytical form expressed in terms of the parameters of the DDE in (1.2), i.e., a, a_d and h. One can explicitly determine how these parameters are involved in the solution and,

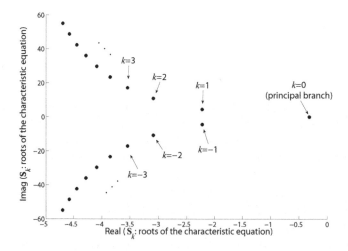

Fig. 2.1   Eigenspectrum of the system (1.2) when $a = -1$, $a_d = 0.5$, and $h = 1$: due to the delay term, $a_d x(t-h)$, and, thus, an exponential term in the characteristic equation, the number of eigenvalues is infinite. The Lambert W function-based approach provides a tool for analysis and control of time-delay systems: each eigenvalue can be expressed analytically in terms of parameters a, a_d, and h, and associated individually with a particular 'branch' ($k = -\infty, \cdots, -1, 0, 1, \cdots, \infty$) of the Lambert W function.

furthermore, how each parameter affects each eigenvalue and the solution. Also, each eigenvalue is associated with k, which indicates the branch of the Lambert W function (see Fig. 2.1). Such an approach has been applied to control problems (Hovel and Scholl, 2005; Wang and Hu, 2008) and extended to other cases, such as fractional-order systems (Chen and Moore, 2002a; Cheng and Hwang, 2006; Hwang and Cheng, 2005) and some special cases of systems of DDEs (Chen and Moore, 2002b; Shinozaki and Mori, 2006; Jarlebring and Damm, 2007).

In this chapter, this analytical approach is extended to general systems of DDEs, including nonhomogeneous DDEs where external inputs are nonzero, and compared with the results obtained by numerical integration. The form of the solution obtained is analogous to the general solution form for ordinary differential equations, and the concept of the state transition matrix in ODEs can be generalized to DDEs using the matrix Lambert W function (see Table 2.2).

2.2 Free Systems of DDEs

2.2.1 *Generalization to free systems of DDEs*

Consider the system of DDEs in matrix-vector form,

$$
\begin{aligned}
\dot{\mathbf{x}}(t) &= \mathbf{A}\mathbf{x}(t) + \mathbf{A_d}\mathbf{x}(t - h), \ t > 0 \\
\mathbf{x}(t) &= \mathbf{x}_0, & t = 0 \\
\mathbf{x}(t) &= \mathbf{g}(t), & t \in [-h, 0)
\end{aligned}
\tag{2.1}
$$

where $\mathbf{A}$ and $\mathbf{A_d}$ are $n \times n$ matrices, and $\mathbf{x}(t)$ is an $n \times 1$ state vector, and $\mathbf{g}(t)$ and $\mathbf{x}_0$ are a specified preshape function and an initial state defined in the Banach space, respectively. For this system of linear DDEs, Hale and Lunel proved the existence and uniqueness of the solution (Hale and Lunel, 1993). In the special case where the coefficient matrices, $\mathbf{A}$ and $\mathbf{A_d}$, commute the solution is given as (Asl and Ulsoy, 2003)

$$
\mathbf{x}(t) = \sum_{k=-\infty}^{\infty} e^{(\frac{1}{h}\mathbf{W}_k(\mathbf{A_d}he^{-\mathbf{A}h}) + \mathbf{A})t} \mathbf{C}_k^I
\tag{2.2}
$$

However, this solution, which is of the same form as the scalar case in Eq. (1.3), is only valid when the matrices $\mathbf{A}$ and $\mathbf{A_d}$ commute, that is $\mathbf{A}\mathbf{A_d} = \mathbf{A_d}\mathbf{A}$ (Yi and Ulsoy, 2006). Therefore, Eq. (2.2) cannot be used for general systems of DDEs and, thus, the solution in Eq. (2.2) is not correct in general. This has been, also, pointed out independently in (Zafer, 2007), (Jarlebring and Damm, 2007) and (Asl and Ulsoy, 2007). The solution in terms of the matrix Lambert W function to systems of DDEs in Eq. (2.1) for the general case is derived here (Yi and Ulsoy, 2006).

First a solution form for Eq. (2.1) is assumed as

$$\mathbf{x}(t) = e^{\mathbf{S}t}\mathbf{C}^I \tag{2.3}$$

where $\mathbf{S}$ is $n \times n$ matrix and $\mathbf{C}^I$ is constant $n \times 1$ vector. Typically, the characteristic equation for (2.1) is obtained by assuming a nontrivial solution of the form $e^{st}\mathbf{C}$ where s is a scalar variable and $\mathbf{C}^I$ is constant $n \times 1$ vector (Hale and Lunel, 1993). Alternatively, one can assume the form of Eq. (2.3) to derive the solution to systems of DDEs in Eq. (2.1) using the matrix Lambert W function. Substitution of Eq. (2.3) into Eq. (2.1) yields,

$$\mathbf{S}e^{\mathbf{S}t}\mathbf{C}^I - \mathbf{A}e^{\mathbf{S}t}\mathbf{C}^I - \mathbf{A_d}e^{\mathbf{S}(t-h)}\mathbf{C}^I = \mathbf{0} \tag{2.4}$$

and

$$\begin{aligned}\mathbf{S}e^{\mathbf{S}t}\mathbf{C}^I - \mathbf{A}e^{\mathbf{S}t}\mathbf{C}^I - \mathbf{A_d}e^{-\mathbf{S}h}e^{\mathbf{S}t}\mathbf{C}^I \\ = (\mathbf{S} - \mathbf{A} - \mathbf{A_d}e^{-\mathbf{S}h})e^{\mathbf{S}t}\mathbf{C}^I = \mathbf{0}\end{aligned} \tag{2.5}$$

Because the matrix $\mathbf{S}$ is an inherent characteristic of a system, and independent of initial conditions, it can be concluded for Eq. (2.5) to be satisfied for any arbitrary initial condition and for every time, t,

$$\mathbf{S} - \mathbf{A} - \mathbf{A_d}e^{-\mathbf{S}h} = \mathbf{0} \tag{2.6}$$

In the special case where $\mathbf{A_d} = \mathbf{0}$, the delay term in Eq. (2.1) disappears, and Eq. (2.1) becomes a system of ODEs, and Eq. (2.6) reduces to

$$\mathbf{S} - \mathbf{A} = \mathbf{0} \Longleftrightarrow \mathbf{S} = \mathbf{A} \tag{2.7}$$

Substitution of Eq. (2.7) into Eq. (2.3), which becomes a system of ODEs only with $\mathbf{x}_0$ without $\mathbf{g}(t)$ (i.e., $\mathbf{C}^I = \mathbf{x}_0$), yields

$$\mathbf{x}(t) = e^{\mathbf{A}t}\mathbf{x}_0 \tag{2.8}$$

This is the well-known solution to a homogeneous system of ODEs in terms of the matrix exponential. Returning to the system of DDEs in Eq. (2.1), one can multiply through by $he^{\mathbf{S}h}e^{-\mathbf{A}h}$ on both sides of Eq. (2.6) and rearrange to obtain,

$$h(\mathbf{S} - \mathbf{A})e^{\mathbf{S}h}e^{-\mathbf{A}h} = \mathbf{A_d}he^{-\mathbf{A}h} \tag{2.9}$$

In general, $\mathbf{S}$ and $\mathbf{A}$ do not commute. It is shown in Appendix A that when $\mathbf{A}$ and $\mathbf{A_d}$ commute, then $\mathbf{S}$ and $\mathbf{A_d}$ also commute. However, in general, $\mathbf{A}$ and $\mathbf{A_d}$ do not commute, and

$$h(\mathbf{S} - \mathbf{A})e^{\mathbf{S}h}e^{-\mathbf{A}h} \neq h(\mathbf{S} - \mathbf{A})e^{(\mathbf{S}-\mathbf{A})h} \qquad (2.10)$$

Consequently, to compensate for the inequality in Eq. (2.10) and to use the *matrix* Lambert W function defined as

$$\mathbf{W}(\mathbf{H})e^{\mathbf{W}(\mathbf{H})} = \mathbf{H} \qquad (2.11)$$

here, an unknown matrix $\mathbf{Q}$ is introduced to satisfy

$$h(\mathbf{S} - \mathbf{A})e^{(\mathbf{S}-\mathbf{A})h} = \mathbf{A_d}h\mathbf{Q} \qquad (2.12)$$

Comparing Eqs. (2.11) and (2.12) yields

$$(\mathbf{S} - \mathbf{A})h = \mathbf{W}(\mathbf{A_d}h\mathbf{Q}) \qquad (2.13)$$

Then the solution matrix, $\mathbf{S}$, is obtained by solving (2.13):

$$\mathbf{S} = \frac{1}{h}\mathbf{W}(\mathbf{A_d}h\mathbf{Q}) + \mathbf{A} \qquad (2.14)$$

Substituting Eq. (2.14) into Eq. (2.9) yields the following condition which can be used to solve for the unknown matrix $\mathbf{Q}$:

$$\mathbf{W}(\mathbf{A_d}h\mathbf{Q})e^{\mathbf{W}(\mathbf{A_d}h\mathbf{Q})+\mathbf{A}h} = \mathbf{A_d}h \qquad (2.15)$$

The matrix Lambert W function, $\mathbf{W}(\mathbf{H})$, is complex valued, with a complex argument $\mathbf{H}$, and has an infinite number of branches $\mathbf{W}_k(\mathbf{H}_k)$, where $k = -\infty, \cdots, -1, 0, 1, \cdots, \infty$ (Asl and Ulsoy, 2003). Corresponding to each branch, k, of the Lambert W function, $\mathbf{W}_k$, there is a solution $\mathbf{Q}_k$ from (2.15), and for $\mathbf{H}_k = \mathbf{A_d}h\mathbf{Q}_k$, the Jordan canonical form $\mathbf{J}_k$ is computed from $\mathbf{H}_k = \mathbf{Z}_k\mathbf{J}_k\mathbf{Z}_k^{-1}$. $\mathbf{J}_k = \mathrm{diag}(J_{k1}(\hat{\lambda}_1), J_{k2}(\hat{\lambda}_2), \ldots, J_{kp}(\hat{\lambda}_p))$, where $J_{ki}(\hat{\lambda}_i)$ is $m \times m$ Jordan block and m is multiplicity of the eigenvalue $\hat{\lambda}_i$. Then, the matrix Lambert W function can be computed as (Pease, 1965)

$$\mathbf{W}_k(\mathbf{H}_k) = \mathbf{Z}_k \left\{ \mathrm{diag}\left(\mathbf{W}_k(J_{k1}(\hat{\lambda}_1)), \ldots, \mathbf{W}_k(J_{kp}(\hat{\lambda}_p))\right) \right\} \mathbf{Z}_k^{-1} \qquad (2.16)$$

where

$$\mathbf{W}_k(J_{ki}(\hat{\lambda}_i)) = \begin{bmatrix} W_k(\hat{\lambda}_i) & W_k'(\hat{\lambda}_i) & \cdots & \frac{1}{(m-1)!}W_k^{(m-1)}(\hat{\lambda}_i) \\ 0 & W_k(\hat{\lambda}_i) & \cdots & \frac{1}{(m-2)!}W_k^{(m-2)}(\hat{\lambda}_i) \\ \vdots & \vdots & \ddots & \vdots \\ 0 & 0 & \cdots & W_k(\hat{\lambda}_i) \end{bmatrix} \qquad (2.17)$$

With the matrix Lambert W function, $\mathbf{W}_k$, given in Eq. (2.16), $\mathbf{S}_k$ is computed from Eq. (2.14). The principal ($k = 0$) and other ($k \neq 0$) branches of the Lambert W function can be calculated from a series definition (Corless et al., 1996) or using commands already embedded in various commercial software packages, such as Matlab, Maple, and Mathematica. With $\mathbf{W}_k$, which satisfies

$$\mathbf{W}_k(\mathbf{H}_k)e^{\mathbf{W}_k(\mathbf{H}_k)} = \mathbf{H}_k \qquad (2.18)$$

finally, the $\mathbf{Q}_k$ is obtained from

$$\mathbf{W}_k(\mathbf{A_d}h\mathbf{Q}_k)e^{\mathbf{W}_k(\mathbf{A_d}h\mathbf{Q}_k)+\mathbf{A}h} = \mathbf{A_d}h \qquad (2.19)$$

and the $\mathbf{Q}_k$ obtained can be substituted into Eq. (2.14),

$$\mathbf{S}_k = \frac{1}{h}\mathbf{W}_k(\mathbf{A_d}h\mathbf{Q}_k) + \mathbf{A} \qquad (2.20)$$

and then $\mathbf{S}_k$ into Eq. (2.3) to obtain the free solution to Eq. (2.1),

$$\mathbf{x}(t) = \sum_{k=-\infty}^{\infty} e^{\mathbf{S}_k t}\mathbf{C}_k^I \qquad (2.21)$$

The coefficient $\mathbf{C}_k^I$ in Eq. (2.21) is a function of $\mathbf{A}$, $\mathbf{A_d}$, h and the pre-shape function, $\mathbf{g}(t)$, and the initial state, $\mathbf{x}_0$. The numerical method for computing $\mathbf{C}_k^I$ were developed in (Asl and Ulsoy, 2003), and an analytical method is also presented in Section 2.4. Conditions for convergence of the infinite series in Eq. (2.21) have been studied in (Banks and Manitius, 1975; Bellman and Cooke, 1963; Hale and Lunel, 1993), and (Lunel, 1989). For example, if the coefficient matrix, $\mathbf{A_d}$, is nonsingular, the infinite series converges to the solution. The solution to DDEs in terms of the Lambert W function, and its analogy to that of ODEs, is summarized in Table 2.2. The matrix $\mathbf{Q}_k$ is obtained numerically from Eq. (2.19), for a variety of initial conditions, for example, using the *fsolve* function in Matlab. In the examples, which have been studied, Eq. (2.19) has a unique solution, $\mathbf{Q}_k$, for each branch, k, if $\mathbf{A_d}$ is nonsingular. When $\mathbf{A_d}$ is rank deficient, some elements of $\mathbf{Q}_k$ do not appear in Eq. (2.19) because they are multiplied by zeros and, thus, are undetermined. In such cases, due to the undetermined elements, $\mathbf{Q}_k$ is clearly not unique. However, those undetermined elements of $\mathbf{Q}_k$ do not appear in $\mathbf{S}_k$ in Eq. (2.20) either, due to multiplication with $\mathbf{A_d}$, and so do not affect the solutions.

Example 2.1. The following example, from (Lee and Dianat, 1981), illustrates the approach and compares the results to those obtained using

Table 2.1   Intermediate results for computing the solution for the example in (2.22) via the matrix Lambert W function.

	$k = 0$	$k = \pm 1$	$\cdots$
$\mathbf{Q}_k$	$\begin{bmatrix} -9.9183 & 14.2985 \\ -32.7746 & 6.5735 \end{bmatrix}$	$\begin{bmatrix} -18.8024 \mp 10.2243i & 6.0782 \mp 2.2661i \\ -61.1342 \mp 23.6812i & 1.0161 \mp 0.2653i \end{bmatrix}$	$\cdots$
$\mathbf{S}_k$	$\begin{bmatrix} 0.3055 & -0.4150 \\ 2.1317 & -3.3015 \end{bmatrix}$	$\begin{bmatrix} -0.3499 \pm 4.9801i & -1.6253 \mp 0.1459i \\ 2.4174 \mp 0.1308i & -5.1048 \pm 4.5592i \end{bmatrix}$	$\cdots$
λ_{ki}	$\begin{cases} -1.0119 \\ -1.9841 \end{cases}$	$\begin{cases} -1.3990 \pm 5.0935i \\ -4.0558 \pm 4.4458i \end{cases}$	$\cdots$

numerical integration. Consider a system of DDEs,

$$\dot{\mathbf{x}}(t) = \begin{bmatrix} -1 & -3 \\ 2 & -5 \end{bmatrix} \mathbf{x}(t) + \begin{bmatrix} 1.66 & -0.697 \\ 0.93 & -0.330 \end{bmatrix} \mathbf{x}(t-1) \qquad (2.22)$$

Then, with the parameters in Eq. (2.22) for solution, $\mathbf{Q}_k$ is computed from Eq. (2.19) for each branch and, subsequently, $\mathbf{S}_k$ is computed from Eq. (2.20). Table 2.1 shows the resulting values for $k = -1, 0, 1$ and the eigenvalues, λ_{k1} and λ_{k2}, of $\mathbf{S}_k$. Using the values of $\mathbf{S}_k$ from Table 2.1, the solution is obtain as

$$\mathbf{x}(t) = \sum_{k=-\infty}^{\infty} e^{\mathbf{S}_k t} \mathbf{C}_k^I = \cdots + e^{\mathbf{S}_{-1} t} \mathbf{C}_{-1}^I + e^{\mathbf{S}_0 t} \mathbf{C}_0^I + e^{\mathbf{S}_1 t} \mathbf{C}_1^I + \cdots \qquad (2.23)$$

The coefficients $\mathbf{C}_k^I$ in (2.23) are determined from specified preshape function, $\mathbf{g}(t)$, initial state, $\mathbf{x}_0$, time delay, h, numerically (Asl and Ulsoy, 2003) or analytically as discussed subsequently in Section 2.4. For example, let $\mathbf{x}_0 = \mathbf{g}(t) = \{1 \ \ 0\}^T$, for $h = 1$, $k = -1, 0, 1$, the corresponding values computed by using the approach in (Asl and Ulsoy, 2003) are

$$\mathbf{C}_{-1}^I = \begin{Bmatrix} 1.3663 + 3.9491i \\ 3.2931 + 9.3999i \end{Bmatrix}, \mathbf{C}_0^I = \begin{Bmatrix} -1.7327 \\ -6.5863 \end{Bmatrix}, \mathbf{C}_1^I = \begin{Bmatrix} 1.3663 - 3.9491i \\ 3.2931 - 9.3999i \end{Bmatrix}$$

$$(2.24)$$

The results are compared to those obtained using numerical integration in Fig. 2.2, and show good agreement as more branches are used. As seen in Fig. 2.2, as one adds terms (i.e., branches), the errors between the two approaches continue to be reduced. However, an explicit expression for the error in terms of the number of branches used is not available.

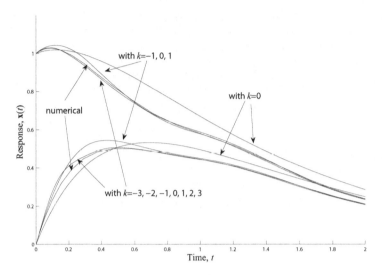

Fig. 2.2   Comparison for example in (2.22) of results from numerical integration vs. (2.23) and (2.24) with one, three, and, seven terms. With more branches the results show better agreement.

2.2.2   *Stability*

For systems of DDEs as in Eq. (2.1), it is difficult to determine the rightmost eigenvalues in the infinite eigenspectrum. However, this is important, as the rightmost eigenvalues determine system stability. If one computes a finite set of eigenvalues from the infinite eigenspectrum, it is difficult to draw a conclusion about stability, because one cannot be sure that the rightmost eigenvalue is included in that finite set. For the scalar case in Eq. (1.2), it has been proven that the root obtained using the principal branch ($k = 0$) always determines the stability of the system using monotinicity of the real part of the Lambert W function with respect to its branch k (Shinozaki and Mori, 2006) (e.g., see Fig. 2.1). Such a proof can readily be extended to systems of DDEs where $\mathbf{A}$ and $\mathbf{A_d}$ are simultaneously triangularizable and, thus, commute with each other (Radjavi and Rosenthal, 2000). Although such a proof is not currently available in the case of the general matrix-vector DDEs in Eq. (2.1), if the coefficient matrix $\mathbf{A_d}$ does not have repeated zero eigenvalues, then, the same behavior has been observed in all the examples which have been have considered. In the example in Eq. (2.22), the value of the real part of the dominant eigenvalue is in the left half plane and, therefore, the system is stable (see Table 2.1).

Consequently, an important advantage of the solution approach based on the Lambert W function, is that the stability of the system can be determined based only on the principal branch. Based on this observation, in Chapter 3, a *Conjecture* (see Subsection 3.3.1) is formulated for stability analysis. Note that for the case where $\mathbf{A_d}$ has repeated zero eigenvalues, it has been observed that the rightmost eigenvalue is obtained by using the principal branch $(k = 0)$, or $k = \pm 1$.

2.3 Forced Systems

Consider a nonhomogeneous version of the DDE in Eq. (1.2):

$$\dot{x}(t) = ax(t) + a_d x(t - h) + bu(t), \, t > 0 \tag{2.25}$$

where $u(t)$ is a continuous function representing the external excitation. In (Malek-Zavarei and Jamshidi, 1987), the forced solution to Eq. (2.25) is presented as,

$$x_{forced}(t) = \int_0^t \Psi(t, \xi) bu(\xi) d\xi \tag{2.26}$$

where the following conditions for the kernel function, $\Psi(t, \xi)$, must be satisfied.

$$a) \frac{\partial}{\partial \xi} \Psi(t, \xi) = -a\Psi(t, \xi), \qquad\qquad t - h \leq \xi < t$$
$$= -a\Psi(t, \xi) - a_d \Psi(t, \xi + h), \, \xi < t - h \tag{2.27}$$
$$b) \Psi(t, t) = 1$$
$$c) \Psi(t, \xi) = 0, \qquad\qquad\qquad \xi > t$$

Because the above conditions contain a scalar DDE, the approach based upon the Lambert W function can be used to obtain $\Psi(t, \xi)$ to extend the free solution in Eq. (1.3) to nonhomogeneous DDE. First, a $\Psi(t, \xi)$ which satisfies the first condition in Eq. (2.27) is

$$\Psi(t, \xi) = e^{a(t-\xi)} \tag{2.28}$$

A $\Psi(t, \xi)$ satisfying the second condition in Eq. (2.27) can be obtained using Eq. (1.3), and can be confirmed by substitution as

$$\Psi(t, \xi) = \sum_{k=-\infty}^{\infty} e^{S_k(t-\xi)} C_k^N \tag{2.29}$$

Thus, it can be concluded that

$$a) \Psi(t, \xi) = e^{a(t-\xi)}, \qquad\qquad t - h \leq \xi < t$$

$$= \sum_{k=-\infty}^{\infty} e^{S_k(t-\xi)} C_k^N, \, \xi < t - h \qquad (2.30)$$

$$b) \Psi(t, \xi) = 0, \qquad\qquad \xi > t$$

Consequently, the forced solution can be represented in terms of the Lambert W function solution as:

Case I $0 \leq t \leq h$

$$x_{forced}(t) = \int_0^t e^{a(t-\xi)} bu(\xi) d\xi \qquad (2.31)$$

Case II $t \geq h$

$$x_{forced}(t) = \int_0^{t-h} \sum_{k=-\infty}^{\infty} e^{S_k(t-\xi)} C_k^N bu(\xi) d\xi + \int_{t-h}^t e^{a(t-\xi)} bu(\xi) d\xi \qquad (2.32)$$

The coefficient, C_k^N, is a function of the parameters of the system in Eq. (2.25), that is, a, a_d and the delay time h. It can be computed approximately depending on the total number of branches, N, used in the solution in a similar way to C_k^I, based on the continuity of Eqs. (2.31) and (2.32) (Yi and Ulsoy, 2006):

$$\underbrace{\left\{ \begin{array}{c} \sigma(h) \\ \sigma(h - \frac{h}{2N}) \\ \sigma(h - \frac{2h}{2N}) \\ \vdots \\ \sigma(0) \end{array} \right\}}_{\bar{\sigma}} = \qquad\qquad (2.33)$$

$$\underbrace{\left[\begin{array}{ccc} \eta_{-N}(h) & \cdots & \eta_N(h) \\ \eta_{-N}(h - \frac{h}{2N}) & \cdots & \eta_N(h - \frac{h}{2N}) \\ \eta_{-N}(h - \frac{2h}{2N}) & \cdots & \eta_N(h - \frac{2h}{2N}) \\ \vdots & \cdots & \vdots \\ \eta_{-N}(0) & \cdots & \eta_N(0) \end{array} \right]}_{\bar{\eta}} \underbrace{\left\{ \begin{array}{c} C_{-N}^N \\ C_{-(N-1)}^N \\ C_{-(N-2)}^N \\ \vdots \\ C_N^N \end{array} \right\}}_{} + \underbrace{\left\{ \begin{array}{c} \pi(h) \\ \pi(h - \frac{h}{2N}) \\ \pi(h - \frac{2h}{2N}) \\ \vdots \\ \pi(0) \end{array} \right\}}_{\bar{\pi}}$$

where

$$\sigma(t) = \int_0^t e^{a(t-\xi)} bu(\xi) d\xi$$

$$\eta_k(t) = \int_0^{t-h} e^{S_k(t-\xi)} bu(\xi) d\xi \tag{2.34}$$

$$\pi(t) = \int_{t-h}^t e^{a(t-\xi)} bu(\xi) d\xi$$

Consequently the C_k^N can be represented as:

$$C_k^N = \lim_{N \to \infty} \lfloor \bar\eta^{-1}(h, N) \cdot (\bar\sigma - \bar\pi) \rfloor_k \tag{2.35}$$

also, C_k^N can be expressed analytically in terms of the system parameters as shown in Section 2.4. The coefficients C_k^I depend on the initial conditions and the preshape function, but as seen from the above procedure, the C_k^N do not. An analytical method to compute C_k^I and C_k^N based on the Laplace transform is presented in Section 2.4. With the obtained C_k^N, using Eqs. (2.33)–(2.35) the forced solution in Eqs. (2.31) and (2.32) is reduced to the forced solution, as shown in Section 2.4 (also alternatively in Appendix A):

$$x_{forced}(t) = \int_0^t \sum_{k=-\infty}^{\infty} e^{S_k(t-\xi)} C_k^N bu(\xi) d\xi \tag{2.36}$$

Hence, combined with Eq. (1.3) the total solution to (2.25) becomes

$$x(t) = \underbrace{\sum_{k=-\infty}^{\infty} e^{S_k t} C_k^I}_{\text{free}} + \underbrace{\int_0^t \sum_{k=-\infty}^{\infty} e^{S_k(t-\xi)} C_k^N bu(\xi) d\xi}_{\text{forced}} \tag{2.37}$$

Example 2.2. Consider Eq. (2.25), with $a = a_d = -1$ and $h = 1$ and the forcing input

$$bu(t) = \cos(t), \, t > 0 \tag{2.38}$$

The total response is shown in Fig. 2.3 for $g(t) = 1$, $x_0 = 1$ with seven branches ($k = -3, -2, -1, 0, 1, 2, 3$), and compared to the result obtained by numerical integration.

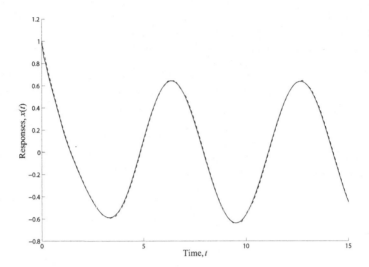

Fig. 2.3   Total forced response with seven branches ($k = -3, -2, -1, 0, 1, 2, 3$) and comparison between the new method (solid) and the numerical method (dashed), which shows good agreement. Parameters are $a = a_d = -1$, $h = 1$ with Eq. (2.38).

2.3.1   *Generalization to systems of DDEs*

The nonhomogeneous matrix form of the delay differential equation in Eq. (2.1) can be written as

$$
\begin{aligned}
\dot{\mathbf{x}}(t) &= \mathbf{A}\mathbf{x}(t) + \mathbf{A_d}\mathbf{x}(t-h) + \mathbf{B}\mathbf{u}(t), && t > 0 \\
\mathbf{x}(t) &= \mathbf{x}_0, && t = 0 \\
\mathbf{x}(t) &= \mathbf{g}(t), && t \in [-h, 0)
\end{aligned}
\tag{2.39}
$$

where $\mathbf{B}$ is an $n \times r$ matrix, and $\mathbf{u}(t)$ is a $r \times 1$ vector. The particular solution can be derived from Eqs. (2.31)–(2.32) as,

Case I $0 \leq t \leq h$

$$
\mathbf{x}_{forced}(t) = \int_0^t e^{\mathbf{A}(t-\xi)} \mathbf{B}\mathbf{u}(\xi) d\xi
\tag{2.40}
$$

Case II $t \geq h$

$$
\mathbf{x}_{forced}(t) = \int_0^{t-h} \sum_{k=-\infty}^{\infty} e^{\mathbf{S}_k(t-\xi)} \mathbf{C}_k^N \mathbf{B}\mathbf{u}(\xi) d\xi + \int_{t-h}^t e^{\mathbf{A}(t-\xi)} \mathbf{B}\mathbf{u}(\xi) d\xi
\tag{2.41}
$$

In Eq. (2.41), $\mathbf{C}_k^N$ is a coefficient matrix of dimension $n \times n$ and can be calculated in the same way as in the scalar case. Like the scalar case in the previous section, Eqs. (2.40)–(2.41) are combined as

$$\mathbf{x}_{forced}(t) = \int_0^t \sum_{k=-\infty}^{\infty} e^{\mathbf{S}_k(t-\xi)} \mathbf{C}_k^N \mathbf{B} \mathbf{u}(\xi) d\xi \tag{2.42}$$

And the total solution is

$$\mathbf{x}(t) = \underbrace{\sum_{k=-\infty}^{\infty} e^{\mathbf{S}_k t} \mathbf{C}_k^I}_{\text{free}} + \underbrace{\int_0^t \sum_{k=-\infty}^{\infty} e^{\mathbf{S}_k(t-\xi)} \mathbf{C}_k^N \mathbf{B} \mathbf{u}(\xi) d\xi}_{\text{forced}} \tag{2.43}$$

where the coefficient $\mathbf{C}_k^I$ in Eq. (2.43) is a function of $\mathbf{A}$, $\mathbf{A_d}$, h and the preshape function $\mathbf{g}(t)$ and the initial condition $\mathbf{x}_0$, while $\mathbf{C}_k^N$ is a function of $\mathbf{A}$, $\mathbf{A_d}$, h and does not depend on $\mathbf{g}$ or $\mathbf{x}_0$. As seen in Eq. (2.43), the total solution of DDEs using the Lambert W function has a similar form to that of ODEs. (see Table 2.2).

Example 2.3. Consider the system of DDEs in Eq. (2.22) with a sinusoidal external excitation:

$$\mathbf{B} \mathbf{u}(t) = \left\{ \begin{array}{c} \cos(t) \\ 0 \end{array} \right\}, t > 0 \tag{2.44}$$

Then the solution to Eq. (2.44), with the same preshape function and initial state, is obtained from Eq. (2.43) and shown in Fig. 2.4. The differences between our new method with seven branches and numerical integration are essentially indistinguishable.

2.4 Approach Using the Laplace Transformation

In this section, solutions to DDEs in the Laplace domain are considered. Transformed DDEs and their solution are compared with the solutions in the time domain as Eq. (2.37) and Eq. (2.43), and the analytical expressions of C_k^I, C_k^N, $\mathbf{C}_k^I$ and $\mathbf{C}_k^N$ are obtained in terms of system parameters by using the Lambert W function.

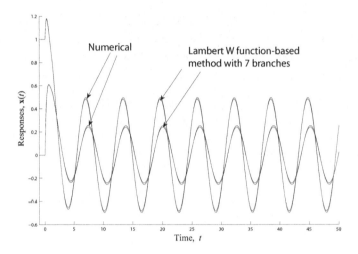

Fig. 2.4   Total response for (2.44) and a comparison of the new method with numerical integration.

2.4.1   *Scalar case*

Consider the scalar free DDE in Eq. (1.2). The Laplace transform of the free equation is

$$sX(s) - x_0 - a_d e^{-sh} X(s) - a_d e^{-sh} G(s) - aX(s)$$
$$= (s - a_d e^{-sh} - a)X(s) - x_0 - a_d G(s) = 0 \tag{2.45}$$

Then,

$$X(s) = \frac{x_0 + a_d e^{-sh} G(s)}{s - a_d e^{-sh} - a} \tag{2.46}$$

On the other hand, the solution obtained by the approach using the Lambert W function in Eq. (1.3) can be transformed as

$$X(s) = \cdots + \frac{C_{-1}^I}{s - S_{-1}} + \frac{C_0^I}{s - S_0} + \frac{C_1^I}{s - S_1} + \cdots$$
$$= \sum_{k=-\infty}^{\infty} \frac{C_k^I}{s - S_k} \tag{2.47}$$

where S_k is obtained from Eq. (1.3). Two solutions in Eqs. (2.46) and (2.47) are compared to derive C_k^I analytically. After some algebraic manipulation and using L'Hopital's rule, S_k is substituted into both equations to get (Yi *et al.*, 2006b)

$$C_k^I = \frac{x_0 + a_d e^{-S_k h} G(S_k)}{1 + a_d h e^{-S_k h}} \tag{2.48}$$

For the nonhomogeneous DDE in Eq. (2.25), C_k^N is also obtained in the same way as

$$C_k^N = \frac{1}{1 + a_d h e^{-S_k h}} \tag{2.49}$$

Note that C_k^I is dependent on the initial conditions, x_0 and the preshape function, $g(t)$, but C_k^N is not. As seen in Eq. (2.48) and Eq. (2.49), comparing the solution in the Laplace domain and that in the time domain using the Lambert W function enables one to derive the analytical expressions for C_k^I and C_k^N. Thus, with S_k in Eq. (1.3), the solution in Eq. (2.37) is explicitly expressed in terms of parameters of the scalar DDE, a, a_d and h.

2.4.2 *Generalization to systems of DDEs*

For the system of DDEs in Eq. (2.39), if one takes the Laplace transform, the unknown $\mathbf{X}(s)$ yields, as in the scalar case in Eq. (2.45),

$$\mathbf{x}(t) = \underbrace{L^{-1}\left[(s\mathbf{I} - \mathbf{A} - \mathbf{A_d}e^{-sh})^{-1}\{\mathbf{x_0} + \mathbf{A_d}e^{-sh}\mathbf{G}(s)\}\right]}_{\text{free}}$$

$$+ \underbrace{L^{-1}\left[(s\mathbf{I} - \mathbf{A} - \mathbf{A_d}e^{-sh})^{-1}\{\mathbf{B}\mathbf{U}(s)\}\right]}_{\text{forced}} \tag{2.50}$$

On the other hand, the free solution to Eq. (2.1) is Eq. (2.21) and it can be transformed as

$$\mathbf{X}(s) = \sum_{k=-\infty}^{\infty} (s\mathbf{I} - \mathbf{S}_k)^{-1}\mathbf{C}_k^I =$$

$$\cdots + (s\mathbf{I} - \mathbf{S}_{-1})^{-1}\mathbf{C}_{-1}^I + (s\mathbf{I} - \mathbf{S}_0)^{-1}\mathbf{C}_0^I + (s\mathbf{I} - \mathbf{S}_1)^{-1}\mathbf{C}_1^I + \cdots \tag{2.51}$$

Comparing Eq. (2.51) with the free solution part in Eq. (2.50) provides the condition for calculating $\mathbf{C}_k^I$. This is analogous to determining the residues in a partial fraction expansion for ODE's. Here, a 2×2 example is provided. If the coefficients are

$$\mathbf{A} = \begin{bmatrix} a_1 & a_2 \\ a_3 & a_4 \end{bmatrix}, \quad \mathbf{A_d} = \begin{bmatrix} a_{d1} & a_{d2} \\ a_{d3} & a_{d4} \end{bmatrix} \tag{2.52}$$

the term in Eq. (2.50) can be written, using the inverse of the matrix, as

$$(s\mathbf{I} - \mathbf{A} - \mathbf{A_d}e^{-sh})^{-1} = \frac{1}{\Upsilon(s)} \begin{bmatrix} s - a_4 - a_{d4}e^{-sh} & a_2 + a_{d2}e^{-sh} \\ a_3 + a_{d3}e^{-sh} & s - a_1 - a_{d1}e^{-sh} \end{bmatrix}$$

(2.53)

where $\Upsilon(s)$ is defined as

$$\Upsilon(s) \equiv s^2 - \{a_1 + a_4 + (a_{d1} + a_{d4})e^{-sh}\}s + (a_1 a_4 - a_2 a_3) +$$
$$(a_1 a_{d4} + a_{d1}a_4 + a_2 a_{d3} + a_{d2}a_3)e^{-sh} + (a_{d1}a_{d4} - a_{d2}a_{d3})e^{-2sh}$$

(2.54)

And the term in (2.51) can be written as

$$s\mathbf{I} - \mathbf{S}_k = \left(s\begin{bmatrix} 1 & 0 \\ 0 & 1 \end{bmatrix} - \begin{bmatrix} p_{k1} & p_{k2} \\ p_{k3} & p_{k4} \end{bmatrix} \right) = \left(s\begin{bmatrix} 1 & 0 \\ 0 & 1 \end{bmatrix} - \mathbf{V}_k \begin{bmatrix} \lambda_{k1} & 0 \\ 0 & \lambda_{k2} \end{bmatrix} \mathbf{V}_k^{-1} \right),$$

$$\text{where } \mathbf{S}_k = \begin{bmatrix} p_{k1} & p_{k2} \\ p_{k3} & p_{k4} \end{bmatrix}$$

(2.55)

Applying Eq. (2.53) and Eq. (2.55), one can find the coefficients $\mathbf{C}_k^I$ in Eq. (2.21). For example, to obtain the coefficient of the principal branch, $\mathbf{C}_0^I$,

$$\frac{1}{\Upsilon(s)} \begin{bmatrix} s - a_4 - a_{d4}e^{-sh} & a_2 + a_{d2}e^{-sh} \\ a_3 + a_{d3}e^{-sh} & s - a_1 - a_{d1}e^{-sh} \end{bmatrix} \times \{\mathbf{x}(0) + \mathbf{A_d}e^{-sh}\mathbf{G}(s)\}$$

$$= \frac{1}{(s - \lambda_{01})(s - \lambda_{02})} \begin{bmatrix} s - p_{04} & p_{02} \\ p_{03} & s - p_{01} \end{bmatrix} + (s\mathbf{I} - \mathbf{S}_{-1})^{-1}\mathbf{C}_{-1}^I$$

$$+ (s\mathbf{I} - \mathbf{S}_1)^{-1}\mathbf{C}_1^I + \cdots$$

(2.56)

Multiply $(s - \lambda_{01})(s - \lambda_{02})$ on both sides to get

$$\frac{(s - \lambda_{01})(s - \lambda_{02})}{\Upsilon(s)} \times \begin{bmatrix} s - a_4 - a_{d4}e^{-sh} & a_2 + a_{d2}e^{-sh} \\ a_3 + a_{d3}e^{-sh} & s - a_1 - a_{d1}e^{-sh} \end{bmatrix}$$

$$\times \{\mathbf{x}(0) + \mathbf{A_d}e^{-sh}\mathbf{G}(s)\} = \begin{bmatrix} s - p_{04} & p_{02} \\ p_{03} & s - p_{01} \end{bmatrix}\mathbf{C}_0^I$$

(2.57)

$$+ (s - \lambda_{01})(s - \lambda_{02})(s\mathbf{I} - \mathbf{S}_{-1})^{-1}\mathbf{C}_{-1}^I$$

$$+ (s - \lambda_{01})(s - \lambda_{02})(s\mathbf{I} - \mathbf{S}_1)^{-1}\mathbf{C}_1^I + \cdots$$

Then, substitution of λ_{01} for s in Eq. (2.57) makes the other terms on the right hand side zero except the first term. And after some algebraic manipulation and using L'Hopital's rule as in the scalar case, one can obtain (Yi *et al.*, 2006b)

$$\lim_{s \to \lambda_{0i}} \frac{\frac{\partial}{\partial s}(s - \lambda_{01})(s - \lambda_{02})}{\frac{\partial}{\partial s}\Upsilon(s)} \times \begin{bmatrix} \lambda_{0i} - a_4 - a_{d4}e^{-sh} & a_2 + a_{d2}e^{-sh} \\ a_3 + a_{d3}e^{-sh} & \lambda_{0i} - a_1 - a_{d1}e^{-sh} \end{bmatrix}$$

$$\times \{\mathbf{x}_0 + \mathbf{A_d}e^{-sh}\mathbf{G}(\lambda_{01})\}$$

$$= \begin{bmatrix} \lambda_{0i} - p_{04} & p_{02} \\ p_{03} & \lambda_{0i} - p_{01} \end{bmatrix} \mathbf{C}_0^I, \quad \text{for } i = 1, 2$$

$$(2.58)$$

and $\mathbf{C}_0^I$, is computed by solving the two equations in Eq. (2.58) simultaneously. Also, for the other branches, $\mathbf{C}_k^I$ is computed with λ_{ki}, where $k = -\infty, \ldots, -1, 1, \ldots, \infty$.

Similarly, the coefficients $\mathbf{C}_k^N$ are computed by comparing the forced parts of (2.43) and (2.50), that is,

$$(s\mathbf{I} - \mathbf{A} - \mathbf{A_d}e^{-sh})^{-1} = \sum_{k=-\infty}^{\infty} (s\mathbf{I} - \mathbf{S}_k)^{-1}\mathbf{C}_k^N$$

$$= \cdots + (s\mathbf{I} - \mathbf{S}_{-1})^{-1}\mathbf{C}_{-1}^N + (s\mathbf{I} - \mathbf{S}_0)^{-1}\mathbf{C}_0^N + \cdots$$

$$(2.59)$$

Then, following a similar derivation, one can get the equation for $\mathbf{C}_k^N$ as

$$\lim_{s \to \lambda_{ki}} \frac{\frac{\partial}{\partial s}(s - \lambda_{k1})(s - \lambda_{k2})}{\frac{\partial}{\partial s}\Upsilon(s)} \times \begin{bmatrix} \lambda_{ki} - a_4 - a_{d4}e^{-sh} & a_2 + a_{d2}e^{-sh} \\ a_3 + a_{d3}e^{-sh} & \lambda_{ki} - a_1 - a_{d1}e^{-sh} \end{bmatrix}$$

$$= \begin{bmatrix} \lambda_{ki} - p_{k4} & p_{k2} \\ p_{k3} & \lambda_{ki} - p_{k1} \end{bmatrix} \mathbf{C}_k^N, \quad \text{for } i = 1, 2$$

$$(2.60)$$

Solving the two equations in Eq. (2.60) simultaneously, one can compute $\mathbf{C}_k^N$. The above approach can readily be generalized to the case of higher order systems of DDEs.

Example 2.4. Consider the example in Eq. (2.22) with an external forcing term of

$$\mathbf{B}u(t) = \begin{bmatrix} cos(t) \\ sin(t) \end{bmatrix}$$

$$(2.61)$$

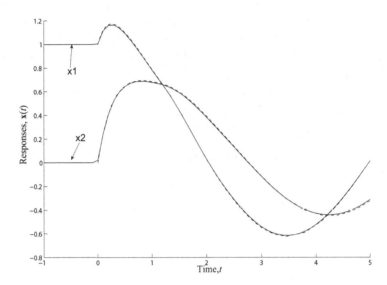

Fig. 2.5   Solution obtained using the Laplace transform combined with the matrix Lambert W function method of 11 branches (straight). Compared to those obtained using the numerical method (dashed), *dde23* in Matlab, they show good agreement.

The coefficient $\mathbf{C}_k^I$ and $\mathbf{C}_k^N$ are computed from Eq. (2.58) and Eq. (2.60), respectively. Applying these values into Eq. (2.43), one can obtain the solution to Eq. (2.22) with Eq. (2.61). The result obtained using 11 branches is shown in Fig. 2.5 and compared to that obtained using the numerical integration method (*dde23* in Matlab). As seen in the figure, the agreement is excellent.

2.5   Concluding Remarks

In this chapter, the Lambert W function-based approach for solution of linear delay differential equations is extended to general systems of DDEs, including nonhomogeneous systems. The solution obtained using the matrix Lambert W function is in a form analogous to the state transition matrix in the solution to systems of linear ordinary differential equations (see Table 2.2). Free and forced responses for several cases of DDEs are presented based on this new solution approach and compared with those obtained by numerical integration. Unlike other solutions to systems of

Table 2.2 Comparison of the solutions to ODEs and DDEs. The solution to DDEs in terms of the Lambert W function shows a formal semblance to that of ODEs.

ODEs

Scalar Case

$$\dot{x}(t) = ax(t) + bu(t), \quad t > 0$$
$$x(t) = x_0, \quad t = 0$$

$$x(t) = e^{at}x_0 + \int_0^t e^{a(t-\xi)}bu(\xi)d\xi$$

Matrix-Vector Case

$$\dot{\mathbf{x}}(t) = \mathbf{A}\mathbf{x}(t) + \mathbf{B}\mathbf{u}(t), \quad t > 0$$
$$\mathbf{x}(t) = \mathbf{x}_0, \quad t = 0$$

$$\mathbf{x}(t) = e^{\mathbf{A}t}\mathbf{x}_0 + \int_0^t e^{\mathbf{A}(t-\xi)}\mathbf{B}\mathbf{u}(\xi)d\xi$$

DDEs

Scalar Case

$$\dot{x}(t) = ax(t) + a_d x(t-h) + bu(t), \quad t > 0$$
$$x(t) = g(t), \quad t \in [-h, 0); \; x(t) = x_0, \quad t = 0$$

$$x(t) = \sum_{k=-\infty}^{\infty} e^{S_k t}C_k^I + \int_0^t \sum_{k=-\infty}^{\infty} e^{S_k(t-\xi)}C_k^N bu(\xi)d\xi$$

$$\text{where, } S_k = \frac{1}{h}W_k(a_d h e^{-ah}) + a$$

Matrix-Vector Case

$$\dot{\mathbf{x}}(t) = \mathbf{A}\mathbf{x}(t) + \mathbf{A_d}\mathbf{x}(t-h) + \mathbf{B}\mathbf{u}(t), \quad t > 0$$
$$\mathbf{x}(t) = \mathbf{g}(t), \quad t \in [-h, 0) \; ; \; \mathbf{x}(t) = \mathbf{x}_0, \quad t = 0$$

$$\mathbf{x}(t) = \sum_{k=-\infty}^{\infty} e^{\mathbf{S}_k t}\mathbf{C}_k^I + \int_0^t \sum_{k=-\infty}^{\infty} e^{\mathbf{S}_k(t-\xi)}\mathbf{C}_k^N \mathbf{B}\mathbf{u}(\xi)d\xi$$

$$\text{where, } \mathbf{S}_k = \frac{1}{h}\mathbf{W}_k(\mathbf{A_d}h\mathbf{Q}_k) + \mathbf{A}$$

DDEs (i.e., Eq. (2.1)) the main contributions of the research presented in this chapter are:

(1) The solution to Eq. (2.1) in Eq. (2.43), in terms of the matrix Lambert W function, is given explicitly in terms of the system coefficients $\mathbf{A}$, $\mathbf{A_d}$, $\mathbf{B}$ and the time delay, h.
(2) Although the eigenspectrum of Eq. (2.1) is infinite, each eigenvalue is distinguished by k, which indicates a branch of the Lambert W function.

(3) If $\mathbf{A_d}$ does not have repeated zero eigenvalues, then, it is our observation that the stability of Eq. (2.1) is determined by the principal branch ($k = 0$).

Even though time-delay systems are still resistant to many methods from control theory (Richard, 2003), the presented approach suggests that some analyses used in systems of ODEs, based on the concept of the state transition matrix, can potentially be extended to systems of DDEs. In systems of ODEs, the parameters of the system appear explicitly in the solutions. Using the Lambert W function, the solution to system of DDEs can be expressed in terms of the coefficients and the delay time, h, as in the ODE case. This approach, with the state transition matrix concept, can pave the way to application of methods from control theory to systems of DDEs, and such an extension is presented in subsequent chapters.

It is noted that there are still several currently outstanding fundamental research problems. First, the method using the matrix Lambert W function hinges on the determination of a matrix, $\mathbf{Q}_k$. As discussed in Section 2.2, it has always been possible to find $\mathbf{Q}_k$ for the problems, which have been considered. However, conditions for the existence and uniqueness of $\mathbf{Q}_k$ are lacking and needed. Second, as discussed in Section 2.2.2, it has been observed in all our examples using DDEs that, when $\mathbf{A_d}$ does not have repeated zero eigenvalues, stability is determined by the principal branch (i.e., $k = 0$) of the matrix Lambert W function. This observation has been proven to be correct in the scalar case and for some special forms of the vector case, however a general proof is lacking. These, and others, are all potential topics for future research, which can build upon the foundation presented in this chapter.

Chapter 3

Stability of Systems of DDEs via the Lambert W Function with Application to Machine Tool Chatter

In a turning process model represented by delay differential equations, the stability of the regenerative machine tool chatter problem is investigated. An approach using the matrix Lambert W function for the analytical solution to systems of delay differential equations, introduced in the previous chapter, is applied to this problem and compared with the result obtained using a bifurcation analysis. The Lambert W function-based approach, known to be useful for solving scalar first-order DDEs, was extended to solve general systems of DDEs in Chapter 2. The essential advantages of the matrix Lambert W function-based approach are not only the similarity to the concept of the state transition matrix in linear ordinary differential equations, enabling its use for general classes of linear delay differential equations, but also the observation that only the finite number of roots obtained by using one branch, the principal branch, among an infinite number of branches is needed to determine the stability of a system of DDEs. With this approach, one can obtain the critical values of delay that determine the stability of a system and hence the preferred operating spindle speed without chatter. In this chapter, the matrix Lambert W function-based approach is applied to the problem of chatter stability in turning, and the result is compared with previous results using existing methods. The new approach shows excellent accuracy and certain other advantages, when compared to traditional graphical, computational and approximate methods.

3.1 Introduction

Machine tool chatter, which can be modeled as a time-delay system, is one of the major constraints that limit the productivity of the turning process. Chatter is the self-excited vibration that is caused by the interaction between the machine structure and the cutting process dynamics. The interaction between the tool-workpiece structure and the cutting process dynamics can be described as a closed-loop system (e.g., see Fig. 3.2). If this system becomes unstable (equivalently, the system of DDEs that represents the process has any unstable eigenvalues), chatter occurs and leads to deteriorated surface finish, dimensional inaccuracy in the machined part, and unexpected damage to the machine tool, including tool breakage. Following the introduction of the classical chatter theories introduced by Tobias (1965) and Tlusty (2000) in the 1960s, various models were developed to predict the onset of chatter. Tobias (1965) developed a graphical method and an algebraic method to determine the onset of instability of a system with multiple degrees of freedom (DOF). Merritt presented a theory to calculate the stability boundary by plotting the harmonic solutions of the system's characteristic equation, assuming that there were no dynamics in the cutting process, and also proposed a simple asymptotic borderline to assure chatter-free performance at all spindle speeds (Merritt, 1965). Optiz and Bernardi (1970) developed a general closed loop representation of the cutting system dynamics for turning and milling processes. The machine structural dynamics were generally expressed in terms of transfer matrices, while the cutting process was limited by two assumptions: (1) direction of the dynamic cutting force is fixed during cutting, and (2) the effects of feed and cutting speed are neglected. These assumptions were later removed by Minis *et al.* (1990), who described the system stability in terms of a characteristic equation and then applied the Nyquist stability criterion to determine the stability of the system. (Chen *et al.*, 1997) introduced a computational method that avoids lengthy algebraic (symbolic) manipulations in solving the characteristic equation. In (Chen *et al.*, 1997), the characteristic equation was numerically formulated as an equation in a single unknown, but well bounded, variable. Also, the stability criteria for time-delay systems were analytically derived by Stepan (1989), Kuang (1993), and Stepan and Moon (1997), and using the Hopf Bifurcation Theorem (Nayfeh *et al.*, 1997; Kalmar-Nagy *et al.*, 2001; Fofana, 2003). Recently, Olgac and Sipahi developed an approach based on the cluster treatment of characteristic roots, examining one infinite cluster of roots at a time for stability of delay systems to enable the determination of the complete stability regions of delay (Sipahi and Olgac, 2003b), and also applied the approach to machining chatter (Olgac and Sipahi, 2005).

In this chapter, an approach based on the matrix Lambert W function for the problem of chatter stability by solving the chatter equation is presented. By applying the approach in Chapter 2 to the chatter equation, one can solve systems of DDEs in the time domain and determine the stability of the system from the eigenvalues in terms of the Lambert W function. Using this method one can obtain ranges of preferred operating spindle speed that does not cause chatter. The form of the solution obtained is

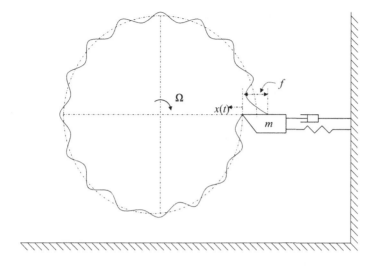

Fig. 3.1   1 DOF orthogonal cutting model.

analogous to the general solution form for ordinary differential equations (ODEs), and the concept of the state transition matrix in ODEs can be generalized to DDEs with the presented method.

3.2   The Chatter Equation in the Turning Process

In the turning process, a cylindrical workpiece rotates with a constant angular velocity, and the tool generates a surface as material is removed. Any vibration of the tool is reflected on this surface, which means that the cutting force depends on the position of the tool edge for the current revolution as well as the previous one, which is reflected on the surface. Thus, to represent such a phenomenon, delay differential equations have been widely used as models for regenerative machine tool vibration. The model of tool vibration, assuming a 1-DOF orthogonal cutting depicted in Fig. 3.1, can be expressed as (Kalmar-Nagy *et al.*, 2001)

$$
\begin{aligned}
&\ddot{x}(t) + 2\zeta\omega_n\dot{x}(t) + \left(\omega_n^2 + \frac{k_c}{m}\right)x(t) - \frac{k_c}{m}x(t-T) \\
&= \frac{k_c}{8f_0 m}\left((x(t) - x(t-T))^2 - \frac{5}{12f_0}(x(t) - x(t-T))^3\right),
\end{aligned}
\tag{3.1}
$$

where $x(t)$ is the general coordinate of tool edge position and the delay, $T = 2\pi/\Omega$, is the time period for one revolution, with Ω being the angular velocity of the rotating workpiece. The coefficient, k_c, is the cutting coefficient derived from a stationary cutting force model as an empirical function of the parameters such as the chip width, the chip thickness, f (nominally f_0 at steady-state), and the cutting speed. The natural angular frequency of the undamped free oscillating system is ω_n and ζ is the relative damping factor. Note that the zero value of the general coordinate $x(t)$ of the tool edge position is selected such that the x component of the cutting force is in balance with the stiffness when the chip thickness, f, is at the nominal value, f_0 (Kalmar-Nagy *et al.*, 2001).

To linearize Eq. (3.1), define $x_1 = x$ and $x_2 \equiv \dot{x}$, and rewrite the equation in first-order form as

$$\dot{x}_1 = x_2(t),$$
$$\dot{x}_2 = -2\zeta\omega_n x_2(t) - \left(\omega_n^2 + \frac{k_c}{m}\right)x_1(t) - \frac{k_c}{m}x_1(t-T)$$
$$+ \frac{k_c}{8f_0m}\left((x_1(t) - x_1(t-T))^2 - \frac{5}{12f_0}(x_1(t) - x_1(t-T))^3\right).$$

$$(3.2)$$

At equilibrium, the condition, $\dot{x}_1(t) = \dot{x}_2(t) = 0$, is satisfied and, thus, the equation becomes

$$0 = x_2(t),$$
$$0 = -2\zeta\omega_n x_2(t) - \left(\omega_n^2 + \frac{k_c}{m}\right)x_1(t) - \frac{k_c}{m}x_1(t-T)$$
$$+ \frac{k_c}{8f_0m}\left((x_1(t) - x_1(t-T))^2 - \frac{5}{12f_0}(x_1(t) - x_1(t-T))^3\right).$$

$$(3.3)$$

and if no vibration from previous processing is left, then $x_1(t) = x_1(t-T) = 0$. Therefore, it can be concluded that one of the equilibrium points is

$$\bar{x}_1(t) = \bar{x}_1(t-T) = \bar{x}_2(t) = 0, \qquad (3.4)$$

which means that at this equilibrium point, the tool edge is in the zero position as defined previously. Linearizing Eq. (3.2) using a Jacobian matrix evaluated at the equilibrium point gives

$$\left\{ \begin{array}{c} \dot{x}_1(t) \\ \dot{x}_2(t) \end{array} \right\} =$$

$$\begin{bmatrix} \dfrac{\partial f}{\partial x_1(t)} & \dfrac{\partial f}{\partial x_2(t)} \\ \dfrac{\partial g}{\partial x_1(t)} & \dfrac{\partial g}{\partial x_2(t)} \end{bmatrix}_0 \left\{ \begin{array}{c} x_1(t) \\ x_2(t) \end{array} \right\} + \begin{bmatrix} \dfrac{\partial f}{\partial x_1(t-T)} & \dfrac{\partial f}{\partial x_2(t-T)} \\ \dfrac{\partial g}{\partial x_1(t-T)} & \dfrac{\partial g}{\partial x_2(t-T)} \end{bmatrix}_0 \left\{ \begin{array}{c} x_1(t-T) \\ x_2(t-T) \end{array} \right\}$$

$$\text{where} \begin{cases} f = x_2 \\ y = -2\zeta\omega_n x_2(t) - \left(\omega_n^2 + \dfrac{k_c}{m} \right) x_1(t) + \dfrac{k_c}{m} x_1(t-T) \\ \quad + \dfrac{k_c}{8f_0 m} \left((x_1(t) - x_1(t-T))^2 - \dfrac{5}{12f_0} (x_1(t) - x_1(t-T))^3 \right) \end{cases}$$

(3.5)

Consider the equilibrium point in Eq. (3.4), Eq. (3.5) becomes

$$\left\{ \begin{array}{c} \dot{x}_1 \\ \dot{x}_2 \end{array} \right\} = \begin{bmatrix} 0 & 1 \\ -\left(\omega_n^2 + \dfrac{k_c}{m} \right) & -2\zeta\omega_n \end{bmatrix} \left\{ \begin{array}{c} x_1(t) \\ x_2(t) \end{array} \right\} + \begin{bmatrix} 0 & 0 \\ \dfrac{k_c}{m} & 0 \end{bmatrix} \left\{ \begin{array}{c} x_1(t-T) \\ x_2(t-T) \end{array} \right\}.$$

(3.6)

Equivalently, Eq. (3.6) can be written as

$$\ddot{x}(t) + 2\zeta\omega_n \dot{x}(t) + \left(\omega_n^2 + \frac{k_c}{m} \right) x(t) - \frac{k_c}{m} x(t-T) = 0 \qquad (3.7)$$

or in the form of (Chen *et al.*, 1997)

$$\frac{1}{\omega_n^2} \ddot{x}(t) + \frac{2\zeta}{\omega_n} \dot{x}(t) + x(t) = -\frac{k_c}{k_m} \left(x(t) - x(t-T) \right), \qquad (3.8)$$

where k_m is structural stiffness (N/m) and $m\omega_n^2 \equiv k_m$.

Figure 3.2 shows the block diagram of the chatter loop. In the diagram, two feedback paths exist: a negative feedback of position (primary path) and a positive feedback of delayed position (regenerative path). The $u_0(s)$ is the nominal depth of cut initially set to zero (Merritt, 1965). Chatter occurs when this closed loop system becomes unstable and, thus, Eq. (3.8) has any unstable eigenvalues. Therefore, the stability of the linearized model in Eq. (3.8) can be used to determine the conditions for the onset of chatter. However, the linearized equations do not capture the amplitude limiting nonlinearities associated with the chatter vibrations. Although comparison with experimental data is not provided here, similar models have been extensively studied and validated in prior works (e.g., see (Chen *et al.*, 1997) and the references therein).

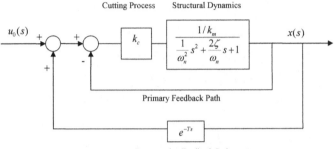

Fig. 3.2   Block diagram of chatter loop (Merritt, 1965). Two feedback paths exist: a negative feedback of position (primary path) and a positive feedback of delayed position (regenerative path). Chatter occurs when this closed loop system becomes unstable.

3.3   Solving DDEs and Stability

The linearized chatter equation (3.8) can be expressed in state space form as Eq. (2.1). Defining $\mathbf{x} = \{x \quad \dot{x}\}^T$, where T indicates transpose, Eq. (3.8) can be expressed as

$$\mathbf{A} = \begin{bmatrix} 0 & 1 \\ -\left(1 + \dfrac{k_c}{k_m}\right)\omega_n^2 & -2\zeta\omega_n \end{bmatrix}, \quad \mathbf{A_d} = \begin{bmatrix} 0 & 0 \\ \dfrac{k_c}{k_m}\omega_n^2 & 0 \end{bmatrix}, \text{and } h = T. \quad (3.9)$$

$\mathbf{A}$ and $\mathbf{A_d}$ are the linearized coefficient matrices of the process model and are functions of the machine-tool and workpiece structural parameters such as natural frequency, damping, and stiffness. The analytical method to solve scalar DDEs, and systems of DDEs as in Eq. (2.1) using the matrix Lambert W function was introduced in the previous chapter. Here the matrix Lambert W function-based approach introduced in Chapter 2 is applied to the chatter problem to find stable operating conditions (e.g., spindle speed and depth of cut). Assume the unknown $\mathbf{Q}$ in Eq. (2.15) as

$$\mathbf{Q} = \begin{bmatrix} q_{11} & q_{12} \\ q_{21} & q_{22} \end{bmatrix}. \quad (3.10)$$

Then matrices in Eqs. (3.9) and (3.10), the argument of the Lambert W function, "$\mathbf{A_d}h\mathbf{Q}$" is

$$\mathbf{A_d}h\mathbf{Q} = \begin{bmatrix} 0 & 0 \\ q_{11}\dfrac{k_c}{k_m}\omega_n^2 T & q_{12}\dfrac{k_c}{k_m}\omega_n^2 T \end{bmatrix}. \quad (3.11)$$

Hence, the eigenvalue matrix and the eigenvector matrix for $\mathbf{A_d}h\mathbf{Q}$ are

$$
\mathbf{d} = \begin{bmatrix} \hat{\lambda}_1 & 0 \\ 0 & \hat{\lambda}_2 \end{bmatrix} = \begin{bmatrix} q_{12}\dfrac{k_c}{k_m}\omega_n^2 T & 0 \\ 0 & 0 \end{bmatrix}, \quad \mathbf{V} = \begin{bmatrix} 0 & -\dfrac{q_{12}}{q_{11}} \\ 1 & 1 \end{bmatrix}. \tag{3.12}
$$

As seen in Eq. (3.12), one of the eigenvalues is zero. This point makes the chatter equation unusual, because of the following property of the Lambert W function (Corless *et al.*, 1996):

$$
W_k(0) = \begin{cases} 0 & \text{when } k = 0 \\ -\infty & \text{when } k \neq 0 \end{cases} \tag{3.13}
$$

Because of this property, in contrast to the typical case where identical branches ($k_1 = k_2$) are used in Eq. (2.16) of the previous chapter, here it is necessary to use hybrid branches ($k_1 \neq k_2$) of the matrix Lambert W function defined as

$$
\mathbf{W}_{k_1,k_2}(\mathbf{A_d}h\mathbf{Q}) = \mathbf{V} \begin{bmatrix} W_{k_1}\left(q_{12}\dfrac{k_c}{k_m}\omega_n^2 T\right) & 0 \\ 0 & W_{k_2}(0) \end{bmatrix} \mathbf{V}^{-1}. \tag{3.14}
$$

By setting $k_2 = 0$ and varying only k_1 from $-\infty$ to ∞, one can solve Eq. (2.15) to get $\mathbf{Q}_{k_1,0}$; then using Eq. (2.14), one determines the transition matrices of the system (2.1) with the coefficients in Eq. (3.9). The results for gain (k_c/k_m) = 0.25, spindle speed ($1/T$) = 50, $\omega_n = 150(\text{sec}^{-2})$, and $\zeta = 0.05$, are in Table 3.1. As seen in Table 3.1, even though k_1 varies, it is observed that the eigenvalues for $k_1 = k_2 = 0$ repeat, which is caused by the fact that one of the branches (k_2) is always zero.

The responses, obtained by using the approach in Chapter 2 with the transition matrices in Table 3.1, are illustrated in Fig. 3.3 and compared with the response using a numerical integration of the nonlinear equation (3.1) and the linearized one (3.7). Note that this is for the linearized equation given by Eq. (2.1) with the coefficients in Eq. (3.9). As seen in Fig. 3.3, because there are an infinite number of transition matrices for DDEs with varying branches, as more transition matrices are utilized, the response approaches the numerically obtained response.

3.3.1 *Eigenvalues and stability*

The solution approach based on the Lambert W function in Eq. (2.21) reveals that the stability condition for the system (2.1) depends on the eigenvalues of the matrix $\mathbf{S}_k$. That is, a time-delay system characterized

Table 3.1   Results of calculation for the chatter equation.

	S_{k_1,k_2}	Eigenvalues of S_{k_1,k_2}
$k_1 = k_2 = 0$	$\begin{bmatrix} 0 & 1 \\ -33083 & -0.24 \end{bmatrix}$	$\begin{cases} -0.12 + 181.88i \\ -0.12 - 181.88i \end{cases}$
$k_1 = -1$ & $k_2 = 0$	$\begin{bmatrix} 0 & 1 \\ -77988 + 32093i & -177 - 247i \end{bmatrix}$	$\begin{cases} -0.12 + 181.88i \\ -176.73 - 428.66i \end{cases}$
	$\begin{bmatrix} 0 & 1 \\ -11 - 1663i & -92 - 182i \end{bmatrix}$	$\begin{cases} -91.61 \\ -0.12 - 181.88i \end{cases}$
$k_1 = 1$ & $k_2 = 0$	$\begin{bmatrix} 0 & 1 \\ -77988 - 32093i & -177 + 247i \end{bmatrix}$	$\begin{cases} -0.12 - 181.88i \\ -176.73 + 428.66i \end{cases}$
	$\begin{bmatrix} 0 & 1 \\ -11 + 1663i & -92 + 182i \end{bmatrix}$	$\begin{cases} -91.61 \\ -0.12 + 181.88i \end{cases}$
$k_1 = -2$ & $k_2 = 0$	$\begin{bmatrix} 0 & 1 \\ -137360 + 42340i & -230 - 570i \end{bmatrix}$	$\begin{cases} -0.12 + 181.88i \\ -233.30 - 755.05i \end{cases}$
	$\begin{bmatrix} 0 & 1 \\ 77945 - 31297i & -177 - 611i \end{bmatrix}$	$\begin{cases} -0.12 - 181.88i \\ -176.73 - 428.66i \end{cases}$
$k_1 = 2$ & $k_2 = 0$	$\begin{bmatrix} 0 & 1 \\ -137360 - 42340i & -230 + 570i \end{bmatrix}$	$\begin{cases} -0.12 - 181.88i \\ -233.30 + 755.05i \end{cases}$
	$\begin{bmatrix} 0 & 1 \\ 77945 + 31297i & -177 + 611i \end{bmatrix}$	$\begin{cases} -0.12 + 181.88i \\ -176.73 + 428.66i \end{cases}$
$\vdots$	$\vdots$	$\vdots$

by Eq. (2.1) is asymptotically stable if and only if all the eigenvalues of S_k have negative real parts. However, computing the matrices S_k for an infinite number of branches, $k = -\infty, \cdots, -1, 0, 1, \cdots, \infty$, is not practical. As explained in Subsection 2.2.2, if the coefficient matrix, A_d, does not have repeated zero eigenvalues, then, it has been observed that the characteristic roots of Eq. (2.1) obtained using only the principal branch are the rightmost ones in the complex plane and determine the stability of the system in Eq. (2.1). Since there is currently no general proof, these observations are formally summarized here in the form of a *Conjecture*. That is,
Conjecture:

> if A_d does not have repeated zero eigenvalues, then
> $\max\{\Re\{\text{eigenvalues for the principal branch, } k = 0\}\} \geq \qquad (3.15)$
> $\Re\{\text{all other eigenvalues}\}$

Note that if A_d has repeated zero eigenvalues, the rightmost eigenvalues are obtained by using the principal branch ($k = 0$), or $k = \pm 1$, for all cases considered.

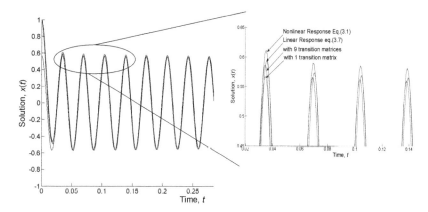

Fig. 3.3 Responses for the chatter equation in Eq. (3.1). With more branches used, the results show better agreement.

The eigenvalues in Table 3.1 are presented in the complex plane in Fig. 3.4. Figure 3.4 shows that the eigenvalues obtained using the principal branch ($k_1 = k_2 = 0$) are closest to the imaginary axis and determine the stability of the system (3.8). For the scalar DDE case, it has been proven that the root obtained using the principal branch always determines stability (Shinozaki and Mori, 2006), and such a proof can readily be extended to systems of DDEs where $\mathbf{A}$ and $\mathbf{A_d}$ commute. However, such a proof is not available in the case of general matrix-vector DDEs. Nevertheless, the same behavior has been observed in all the examples that have been considered. That is, the eigenvalues of $\mathbf{S}_{0,0}$, obtained using the principal branch for both of k_1 and k_2, are closest to the imaginary axis, and their real parts are negative. Furthermore, using additional branches to calculate the eigenvalues always yields eigenvalues whose real parts are further to the left in the s-plane. Thus, it can be concluded that the system (3.8) with the parameter set is stable.

If one observes the roots obtained using the principal branch, one can find the critical point when the roots cross the imaginary axis. For example, when spindle speed ($1/T$) = 50, $\omega_n = 150(\sec^{-2})$ and $\zeta = 0.05$, the critical ratio of gains (k_c/k_m) is 0.2527. This value agrees with the result obtained by the Lyapunov method (Malek-Zavarei and Jamshidi, 1987), the Nyquist criterion and the computational method of (Chen *et al.*, 1997). The stability lobes by this method are depicted in Fig. 3.5 with respect to the spindle speed (*rps*, revolution per second).

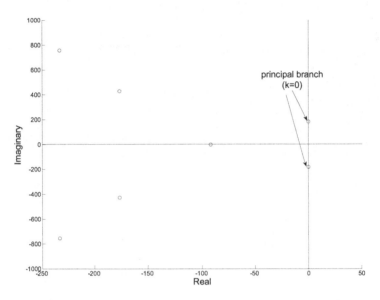

Fig. 3.4   Eigenvalues in Table 3.1 in the complex plane. The eigenvalues obtained using the principal branch ($k = 0$) are dominant and determine the stability of the system.

In obtaining the result shown in the Fig. 3.5, it is noted that the roots obtained using the principal branch always determine stability. One of the advantages of using the matrix Lambert W function over other methods appears to be the observation that the stability of the system can be obtained from only the principal branch among an infinite number of roots. The main advantage of this method is that solution (2.21) in terms of the matrix Lambert W function is similar to that of ODEs. Hence, the concept of the state transition matrix in ODEs can be generalized to DDEs using the matrix Lambert W function. This suggests that the analytical approach using the matrix Lambert W function can be developed for *time-varying* DDEs based on Floquet theory and such study is being currently investigated.

Recently, Forde and Nelson (2004) developed a bifurcation analysis combined with Sturm sequences for determining the stability of delay differential equations. The method simplifies the task of determining the necessary and sufficient conditions for the roots of a quasi-polynomial to have negative real parts, and was applied to a biological system (Forde and Nelson, 2004). For the chatter problem considered here, the bifurcation analysis presented in (Forde and Nelson, 2004) also provides a useful algorithm for

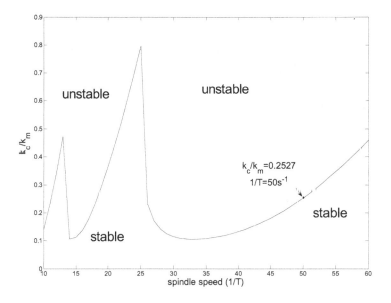

Fig. 3.5   Stability lobes for the chatter equation.

determining stability. In (Yi *et al.*, 2007b), the method was applied to the chatter equation in (3.8). Compared with existing methods, the bifurcation analysis with the Sturm sequence can be used determine the critical values of delay at the stability limit of the system with relatively simple calculations, avoiding restrictive geometric analysis. Also, it showed excellent agreement with the result presented in this chapter.

3.4   Concluding Remarks

In this chapter, a new approach for the stability analysis of machining tool chatter problems, which can be expressed as systems of linear delay differential equations, has been presented using the matrix Lambert W function. The main advantage of the analytical approach based on the matrix Lambert W function lies in the fact that one can obtain the solution to systems of linear DDEs in the time domain, and the solution has a form analogous to the state transition matrix in systems of linear ordinary differential equations. It can be applied to systems of linear DDEs of arbitrary order, and thus can be used in chatter models that include multiple structural vibration modes. Though the solution is in the form of an infinite series of modes computed with different branches, it is observed that the principal branch always determines the stability of a system (e.g., see Fig. 3.4). The results show excellent agreement with those obtained using traditional methods, e.g., Lyapunov (Malek-Zavarei and Jamshidi, 1987), Nyquist, the numerical method used in (Chen *et al.*, 1997), and bifurcation analysis via Sturm sequence (Yi *et al.*, 2007b). The method presented in this chapter not only yields stability results but also can be used to obtain the free and forced response of the linearized machine tool dynamics.

Chapter 4

Controllability and Observability of Systems of Linear Delay Differential Equations via the Matrix Lambert W Function

During recent decades, controllability and observability of linear time-delay systems have been studied, including various definitions and corresponding criteria. However, the lack of an analytical solution approach has limited the applicability of existing theory. Recently, the solution to systems of linear delay differential equations has been derived in the form of an infinite series of modes written in terms of the matrix Lambert W function as introduced in Chapter 2. The solution form enables one to put the results for point-wise controllability and observability of systems of delay differential equations to practical use. In this chapter, the criteria for point-wise controllability and observability are derived, the analytical expressions for their Gramians in terms of the parameters of the system are obtained, and a method to approximate them is developed for the first time using the matrix Lambert W function-based solution form.

4.1 Introduction

The Lambert W function has been used to develop an approach for the solution of linear time invariant (LTI) systems of DDEs with a single delay for scalar first order DDEs and, subsequently, general systems of DDEs (e.g., see Chapter 2 and the references therein). The approach using the Lambert W function provides a solution form for DDEs and thus enables one to put the theoretical results on point-wise controllability and observability of time-delay systems and their Gramians (e.g., see (Malek-Zavarei and Jamshidi, 1987), (Richard, 2003) and the references therein), to practical use. In this chapter, the properties of controllability and observability for time-delay systems are studied via the matrix Lambert W function approach-based solution. Using the analytical solution form in terms of the matrix Lambert W function, the point-wise controllability and observability criteria, and their Gramians, for LTI systems of DDEs with a single delay

43

are derived. Also, the results are applied to an example for illustration. The result provides an analytical approach to investigate the two system input-output properties (controllability and observability), and also is used for obtaining balanced realizations for time-delay systems.

Consider a real LTI system of DDEs with a single constant delay, h, in Eq. (2.39) with an output equation. That is,

$$
\begin{aligned}
\dot{\mathbf{x}}(t) &= \mathbf{A}\mathbf{x}(t) + \mathbf{A_d}\mathbf{x}(t-h) + \mathbf{B}\mathbf{u}(t) \quad t > 0 \\
\mathbf{x}(t) &= \mathbf{g}(t) \qquad\qquad\qquad\qquad\quad t \in [-h, 0) \\
\mathbf{x}(t) &= \mathbf{x}_0 \qquad\qquad\qquad\qquad\quad\; t = 0 \\
\mathbf{y}(t) &= \mathbf{C}\mathbf{x}(t)
\end{aligned}
\tag{4.1}
$$

The coefficient matrix $\mathbf{C}$ is $p \times n$ and $\mathbf{y}(t)$ is a $p \times 1$ measured output vector. Note that there exist two kinds of initial conditions for systems of DDEs, $\mathbf{x}_0$ which is the value of $\mathbf{x}(t)$ at $t = 0$, and the preshape function, $\mathbf{g}(t)$ in Eq. (4.1) and is equal to $\mathbf{x}(t)$ on the interval $t \in [-h, 0)$. For general retarded functional differential equations, the existence and uniqueness of the solution are proved based upon the assumption of continuity, i.e., $\mathbf{g}(0) = \mathbf{x}_0$. However, in the specific case of the LTI system of DDEs with a single constant delay as in Eq. (4.1), the existence and uniqueness can be also proved without such an assumption (Tsoi and Gregson, 1978) and (Hale and Lunel, 1993). Consequently, for generality, one can assume that $\mathbf{g}(0)$ is not necessarily equal to $\mathbf{x}_0$ in Eq. (4.1). In Chapter 2, the solution to Eq. (4.1) was derived using the matrix Lambert W function-based approach and given as in Eq. (2.43).

4.2   Controllability

Controllability and observability are two fundamental attributes of a dynamical system. Such properties of time-delay systems have been explored since the 1960s and the controllability and observability Gramians for time-delay systems were presented respectively by Weiss (1967) and Delfour and Mitter (1972) based upon assumed symbolic solution forms of the DDEs. However, application of the results with Gramians to verify controllability and observability of linear time-delay systems has been difficult, due to the lack of analytical solutions to DDEs (Malek-Zavarei and Jamshidi, 1987). The analysis of controllability and observability based on the solution form in terms of the matrix Lambert W function are presented in this, and subsequent, sections respectively.

Depending on the nature of the problem under consideration, there exist various definitions of controllability and observability for time-delay systems (Richard, 2003) (also see Appendix B for comparison of various types). Among them, the concept of *point-wise controllability* of a system of DDEs, as in Eq. (4.1), and the related conditions were introduced in (Richard, 2003).

Definition 4.1. The system (4.1) is *point-wise controllable* (or equivalently, defined as *fixed-time completely controllable* in (Choudhury, 1972a) or R^n-*controllable to the origin* in (Richard, 2003), (Weiss, 1970)) if, for any given initial conditions $\mathbf{g}(t)$ and $\mathbf{x}_0$, there exists a time t_1, $0 < t_1 < \infty$, and an admissible (i.e., measurable and bounded on a finite time interval) control segment $\mathbf{u}(t)$ for $t \in [0, t_1]$ such that $\mathbf{x}(t_1; \mathbf{g}, \mathbf{x}_0, \mathbf{u}(t)) = \mathbf{0}$ (Weiss, 1967).

The solution form to Eq. (4.1) is assumed as (Bellman and Cooke, 1963)

$$\mathbf{x}(t) \equiv \mathbf{x}(t; \mathbf{g}, \mathbf{x}_0, \mathbf{u}) = \mathbf{M}(t; \mathbf{g}, \mathbf{x}_0) + \int_0^t \mathbf{K}(\xi, t)\mathbf{B}\mathbf{u}(\xi)d\xi, \qquad (4.2)$$

where $\mathbf{M}(t; \mathbf{g}, \mathbf{x}_0)$ is the free solution to Eq. (4.1) and $\mathbf{K}(\xi, t)$ is the kernel function for Eq. (4.1). Then using the kernel $\mathbf{K}(\xi, t)$ in (4.2), the condition for point-wise controllability was derived in (Weiss, 1967) with the following definition.

Definition 4.2. A system (4.1) is *point-wise complete* at time t_1 if, for all $\mathbf{x}_1 \in R^n$, there exist initial conditions $\mathbf{g}(t)$ and $\mathbf{x}_0$, such that $\mathbf{x}(t_1; \mathbf{g}, \mathbf{x}_0, \mathbf{0}) = \mathbf{x}_1$, where $\mathbf{x}(t; \mathbf{g}, \mathbf{x}_0, \mathbf{0})$ is a solution of (4.1) starting at time $t = 0$ (Choudhury, 1972b).

The conditions for point-wise completeness are presented in (Choudhury, 1972b), (Malek-Zavarei and Jamshidi, 1987), and (Thowsen, 1977). For example, all 2×2 DDEs or DDEs with a nonsingular coefficient, $\mathbf{A_d}$, are point-wise complete.

Even though the equations to obtain the kernel function in (4.2) were presented in (Bellman and Cooke, 1963) and (Malek-Zavarei and Jamshidi, 1987) the lack of the knowledge of a solution to the systems of DDEs has prevented the evaluation and application of the results in (Weiss, 1967). This has prompted many authors to develop algebraic controllability criteria in terms of systems matrices (Buckalo, 1968), (Choudhury, 1972a), (Kirillova and Churakova, 1967), and (Weiss, 1970). Other definitions of controllability, which belong in different classifications, such as spectral controllability, have alternatively been provided (Manitius and Olbrot, 1979). For definitions and conditions of various types of controllability and comparisons, refer to (Malek-Zavarei and Jamshidi, 1987), (Richard, 2003) (also see Appendix B).

Using the matrix Lambert W function, however, the linear time-invariant system with a single delay can be solved as in (2.43) and, thus, the kernel function used in the condition for point-wise controllability can be derived. The kernel function $\mathbf{K}(\xi, t_1)$ is obtained, by comparing (4.2) with (2.43), as

$$\mathbf{K}(\xi, t_1) \equiv \sum_{k=-\infty}^{\infty} e^{\mathbf{S}_k(t_1-\xi)} \mathbf{C}_k^N \tag{4.3}$$

Therefore, it is possible to express the controllability Gramian, and state the following main result, for controllability of the systems of DDEs in (4.1)

Theorem 4.1. *If a system (4.1) is point-wise complete, there exists a control which results in point-wise controllability in finite time of the solution of (4.1) for any initial conditions* $g(t)$ *and* $\mathbf{x}_0$, *if and only if*

$$rank\left[C_o(0, t_1) \equiv \int_0^{t_1} \sum_{k=-\infty}^{\infty} e^{\mathbf{S}_k(t_1-\xi)} \mathbf{C}_k^N \mathbf{B}\mathbf{B}^T \left\{ \sum_{k=-\infty}^{\infty} e^{\mathbf{S}_k(t_1-\xi)} \mathbf{C}_k^N \right\}^T d\xi \right] = n \tag{4.4}$$

where $C_o(0, t_1)$ *is the controllability Gramian of the system of DDEs and* ^{T} *indicates the transpose.*

Proof.   *Sufficiency*   In (2.43), in order to transfer $\mathbf{x}(t)$ to $\mathbf{0}$ at t_1, substitute an input obtained with the inverse of the controllability Gramian in (4.4)

$$\mathbf{u}(t) = -\mathbf{B}^T \left\{ \mathbf{K}(t, t_1) \right\}^T C_o^{-1}(0, t_1) \mathbf{M}(t_1; \mathbf{g}, \mathbf{x}_0) \tag{4.5}$$

where $\mathbf{M}$ is the free solution to (4.1), and comparing (4.2) with (2.43) yields

$$\mathbf{M}(t_1; \mathbf{g}, \mathbf{x}_0) \equiv \sum_{k=-\infty}^{\infty} e^{\mathbf{S}_k(t_1-0)} \mathbf{C}_k^I \tag{4.6}$$

then $\mathbf{x}(t_1) = \mathbf{0}$.

Necessity Given any $\mathbf{g}$ and $\mathbf{x}_0$, suppose there exist $t_1 > 0$ and a control $\mathbf{u}_{[0,t_1]}$ such that $\mathbf{x}(t_1) = \mathbf{0}$, but (4.4) does not hold. The latter implies that there exists a non-zero vector $\mathbf{x}_1 \in \Re^n$ such that $\mathbf{x}_1^T \mathbf{K}(t, t_1)\mathbf{B} = \mathbf{0}$, $0 \leq t \leq t_1$ due to the following fact. Let $\mathbf{F}$ be an $n \times p$ matrix. Define

$$P_{(t_1,t_2)} \equiv \int_{t_1}^{t_2} \mathbf{F}(t)\mathbf{F}^T(t)dt \tag{4.7}$$

Then the rows of $\mathbf{F}$ are linearly independent on $[t_1, t_2]$ if and only if the $n \times n$ constant matrix $P_{(t_1,t_2)}$ is nonsingular (Chen, 1984). Then, from (4.2),

$$\mathbf{x}_1^T \mathbf{x}(t_1) = \mathbf{x}_1^T \mathbf{M}(t_1; \mathbf{g}, \mathbf{x}_0) + \int_0^{t_1} \mathbf{x}_1^T \mathbf{K}(\xi, t_1)\mathbf{B}\mathbf{u}(\xi)d\xi \tag{4.8}$$

and $\mathbf{0} = \mathbf{x}_1^T \mathbf{M}(t_1; \mathbf{g}, \mathbf{x}_0)$. By hypothesis, however, $\mathbf{g}$ and $\mathbf{x}_0$ can be chosen such that $\mathbf{M}(t_1; \mathbf{g}, \mathbf{x}_0) = \mathbf{x}_1$. Then $\mathbf{x}_1^T \mathbf{x}_1 = \mathbf{0}$ which contradicts the assumption that $\mathbf{x}_1 \neq \mathbf{0}$. $\qquad\square$

In the ODE case, the input computed using the controllability Gramian will use the minimal energy in transferring $(\mathbf{x}_0, 0)$ to $(\mathbf{0}, t_1)$ (Chen, 1984). Using the controllability Gramian in Eq. (4.4), one can prove that such a result is also available for DDE's in a similar way to the ODE case in (Chen, 1984) (see proof in Appendix B). That is, the input defined in Eq. (4.5) consumes the *minimal* amount of energy, among all the $\mathbf{u}$'s that can transfer $(\mathbf{x}_0, 0)$ to $(\mathbf{0}, t_1)$.

With Theorem 4.1 and Eq. (4.7), assuming that the system (4.1) is point-wise complete, it can be concluded as that

Corollary 4.1. *The system in Eq. (4.1) is point-wise controllable if and only if all rows of*

$$\sum_{k=-\infty}^{\infty} e^{\mathbf{S}_k(t-0)} \mathbf{C}_k^N \mathbf{B} \tag{4.9}$$

are linearly independent on $[0, \infty)$.

The Laplace transform of (4.9) is (Yi *et al.*, 2006b)

$$L\left\{ \sum_{k=-\infty}^{\infty} e^{\mathbf{S}_k(t-0)} \mathbf{C}_k^N \mathbf{B} \right\} = \left(s\mathbf{I} - \mathbf{A} - \mathbf{A}_d e^{-sh} \right)^{-1} \mathbf{B} \tag{4.10}$$

Since the Laplace transform is a one-to-one linear operator, one can obtain the following corollary.

Corollary 4.2. *The system in Eq. (4.1) is point-wise controllable if and only if all rows of*

$$\left(s\boldsymbol{I} - \boldsymbol{A} - \boldsymbol{A}_d e^{-sh}\right)^{-1} \boldsymbol{B} \tag{4.11}$$

are linearly independent, over the field of complex numbers except at the roots of the characteristic equation of Eq. (4.1).

In systems of ODEs, if the state variable $\mathbf{x}(t)$ is forced to zero at $t = t_1$, it stays at zero on $[t_1, \infty)$. However, because the system of DDEs in Eq. (4.1) has a *delayed term* in its equation, even though all the individual state variables are zero at $t = t_1$ they can become non-zero again after t_1. For this reason, additional definitions of controllability for systems of DDEs for functional, not point-wise, types of controllability are available in (Richard, 2003), (Weiss, 1967). Also see Appendix B.

Remark 4.1. It have been shown with some examples in (Yi *et al.*, 2010b) that if the system of DDEs is point-wise controllable, it is possible to design linear feedback controllers via rightmost eigenvalue assignment for the system in Eq. (4.1); otherwise, it is not. This chapter presents the theoretical foundation for establishing point-wise controllability. To date there is no general theory for DDEs, as there is for ODEs, that controllability is required for eigenvalue assignment by linear feedback (Tsoi and Gregson, 1978), (Vandevenne, 1972).

4.3   Observability

Consider the system given in Eq. (4.1). If one knows the initial conditions, $\mathbf{g}(t)$ and $\mathbf{x}_0$, then one can know all state variables for any time using the solution in Eq. (2.43) to the systems of DDEs. As seen in Eq. (2.43), however, the main obstacle is the fact that the free solution does not have the form of just the product of initial conditions and the transition matrix, in contrast to the ODE case. Therefore, a concept of point-wise observability was introduced for systems of DDEs in (Delfour and Mitter, 1972), which is different from that of observability for systems of ODEs.

Definition 4.3. The system in Eq. (4.1) is *point-wise observable*, (or equivalently, *observable* as in (Delfour and Mitter, 1972)) in $[0, t_1]$ if the initial point $\mathbf{x}_0$ can be uniquely determined from the knowledge of $\mathbf{u}(t)$, $\mathbf{g}(t)$, and $\mathbf{y}(t)$ (Delfour and Mitter, 1972).

This concept was introduced by Gabasov *et al.* (1972) for purely mathematical reasons. However, disturbances which can be approximated by Dirac distributions cause the system response to be approximatable by jumps in the trajectory response (Lee and Olbrot, 1981). For such cases, the concept of point-wise observability has been used in analyzing singularly perturbed delay system, where the perturbation is very small but cannot be ignored (see, e.g., (Glizer, 2004; Kopeikina, 1998)).

Just as in the case of controllability, the lack of analytical solutions of the systems of DDEs has prevented the evaluation and application of the above condition. Unlike controllability, the development of algebraic conditions for the investigation of the observability of time delay systems has not received much attention (Malek-Zavarei and Jamshidi, 1987). Bhat and Koivo (1976a) used spectral decomposition to decompose the state space into a finite-dimensional and a complementary part. In (Lee and Olbrot, 1981), various types of observability of time-delay systems and corresponding algebraic conditions were presented. For a detailed study, refer to (Malek-Zavarei and Jamshidi, 1987; Lee and Olbrot, 1981), and the references therein.

Applying the kernel function in Eq. (4.3) to the observability Gramian defined symbolically in (Delfour and Mitter, 1972), one can present the following condition for observability for systems of DDEs. Here the system of (4.1) is assumed to be *point-wise complete*.

Theorem 4.2. *The system in Eq. (4.1) is point-wise observable if and only if*

$$rank\left[O_b(0,t_1) \equiv \int_0^{t_1} \left\{ \sum_{k=-\infty}^{\infty} e^{S_k(\xi-0)} C_k^N \right\}^T C^T C \sum_{k=-\infty}^{\infty} e^{S_k(\xi-0)} C_k^N d\xi \right] = n$$

(4.12)

where $O_b(0,t_1)$ is the observability Gramian of the system of DDEs.

With Theorem 4.2 and Eq. (4.7) in a way similar to controllability study in the previous section, one can conclude that

Corollary 4.3. *The system in Eq. (4.1) is point-wise observable if and only if all columns of the matrix*

$$C \sum_{k=-\infty}^{\infty} e^{S_k(t-0)} C_k^N$$

(4.13)

are linearly independent.

Since the Laplace transform is a one-to-one linear operator, the following corollary then obtained.

Corollary 4.4. *The system in Eq. (4.1) is point-wise observable if and only if all columns of the matrix*

$$C\left(sI - A - A_d e^{-sh}\right)^{-1} \tag{4.14}$$

are linearly independent except at the roots of the characteristic equation of Eq. (4.1).

Proof. The proofs are essentially similar to those of controllability in Section 4.2 and are omitted for brevity. □

Remark 4.2. As in the case of point-wise controllability, for point-wise observable systems of DDEs, a linear asymptotic observer can be designed via rightmost eigenvalue assignment as shown by examples in (Yi *et al.*, 2009).

In the case that $\mathbf{g}(t)$ is unknown, if $\mathbf{g}(t)$, as well as $\mathbf{x}_0$, can be determined uniquely from a knowledge of $\mathbf{u}(t)$ and $\mathbf{y}(t)$, the system of (4.1) is termed *absolutely observable* (or *strongly observable* in (Delfour and Mitter, 1972)). For a detailed explanation of the definition of *absolute observability* and the corresponding conditions, the reader is referred to (Delfour and Mitter, 1972).

4.4   Illustrative Example

Consider a system of DDEs (4.1) with parameters, from (Lee and Dianat, 1981),

$$\mathbf{A} = \begin{bmatrix} -1 & -3 \\ 2 & -5 \end{bmatrix}, \quad \mathbf{A_d} = \begin{bmatrix} 1.66 & -0.697 \\ 0.93 & -0.330 \end{bmatrix}, \quad h = 1 \tag{4.15}$$

The response, using the solution form in Eq. (2.43), is depicted in Fig. 4.1-(a) when $\mathbf{g}(t) = \{\,1\ 0\,\}^T$ and $\mathbf{x}_0 = \{\,1\ 0\,\}^T$. The solution Eq. (2.43) has the form of an infinite series of modes written in terms of the matrix Lambert W function. Even though it is not practically feasible to add all the infinite terms of the series in Eq. (2.43), it can be approximated by a finite number of terms. For example, in Fig. 4.1-(a), 33 branches ($k = -16, \ldots, 16$) of the Lambert W function are used. As one adds terms, the errors between the response (from Eq. (2.43)) and a solution obtained numerically (using

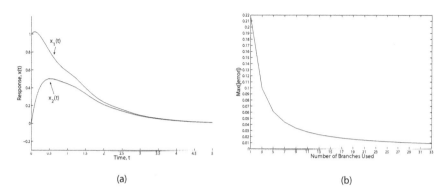

(a)   (b)

Fig. 4.1 Response of the example in Eq. (4.15) obtained by the matrix Lambert W function approach with 33 terms (a) and the maximum of errors between the response by the solution form in Eq. (2.43) and the numerically obtained one (b) corresponding to the number of branches used for the response. The errors continue to be reduced as more terms in the series solution are included.

dde23 in Matlab) continue to be reduced and validate the convergence of the solution in Eq. (2.43) (see Fig. 4.1-(b)).

Using the criterion in (Choudhury, 1972b) (also presented in Section 4.2), the system in Eq. (4.15) is point-wise complete. For $\mathbf{B}=[1\ 0]^T$, the controllability Gramian $C_o(0, t_1)$ in Eq. (4.4) can be computed. Then in order for the system Eq. (4.15) to be point-wise controllable, $C_o(0, t_1)$ should have full rank. This means that the determinant of the matrix is non-zero. That is,

$$\det |C_o(0, t_1)| \neq 0 \qquad (4.16)$$

Computing the determinant of the matrix for an increasing number of branches yields the result in Fig. 4.2. As more branches are included, the solution in Eq. (2.43) converges (see Fig. 4.1), so do the kernel function in (4.3) and the controllability Gramian in Eq. (4.4). Figure 4.2 shows that the determinant converges to a non-zero value, which implies that the system is point-wise controllable.

Even though a system satisfies the algebraic criteria already provided in previous work, such as (Lee and Olbrot, 1981), (Malek-Zavarei and Jamshidi, 1987), in cases where the determinant of the observability Gramian in Eq. (4.12) is smaller than a specific value, then it is not practical to design an observer as the gains in the observer can become unrealistically high. Comparing the determinant of the observability Gramian

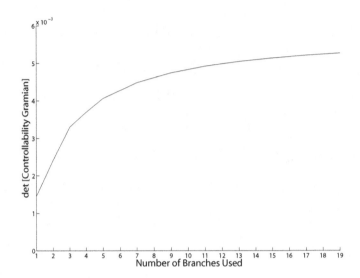

Fig. 4.2   Determinant of the controllability Gramian versus branches. As more branches are included, the value of determinant converges to a non-zero value.

corresponding to the system in Eq. (4.15), one can obtain a practical assessment. For example, the determinants of the observability Gramian for Eq. (4.15) when $\mathbf{C} = [\,1\ 0\,]$ and $\mathbf{C} = [\,0\ 1\,]$ are compared in Fig. 4.3 with $t_1 = 4$. As the number of branches used increases, the value of the determinant in case of $\mathbf{C} = [\,1\ 0\,]$ tends to converge to a higher value than the case of $\mathbf{C} = [\,0\ 1\,]$.

From the results in Figs. 4.2 and 4.3, although a formal study of truncation errors is needed, the convergence of the Gramians is observed as the number of terms in the series is increased. When the convergence conditions, which are explained in Chapter 2, and also in the cited references, are satisfied, then the series expansion of the solution in Eq. (2.43) converges. The controllability Gramian in Eq. (4.4) and the observability Gramian in Eq. (4.12) are the integrals of products of the kernel (Eq. (4.3)) and constant matrices ($\mathbf{B}$ and $\mathbf{C}$) over a finite interval. Thus, the convergence of the Gramians is also assured under the same conditions.

The presented results agree with those obtained using existing algebraic methods. However, using the method of Gramians developed in this paper, one can acquire additional information. The controllability and observability Gramians indicate how controllable and observable the corresponding states are (Holford and Agathoklis, 1996), while the algebraic conditions

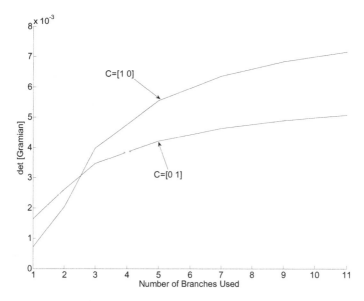

Fig. 4.3 Determinant of observability Gramian when $\mathbf{C} = [0 \; 1]$ and $\mathbf{C} = [1 \; 0]$. As the number of branches used increases, the value of the determinant in case of $\mathbf{C} = [1 \; 0]$ tends to converge to a higher value than for $\mathbf{C} = [0 \; 1]$.

for controllability/observability reveal only whether a system is controllable/observable or not. Therefore, with the conditions using Gramian concepts, one can determine how the change in some specific parameters of the system or the delay time, h, affects the controllability and observability of the system via the changes in the Gramians. Using the Gramians presented in the previous sections, the concept of the balanced realization, in which the controllability Gramian and observability Gramian of a system are equal and diagonal, can be extended to systems of DDEs. For systems of ODEs, the balanced realization has been studied because of its desirable properties such as good error bounds, computational simplicity, stability, and its close connection to robust multi-variable control (Verriest and Kailath, 1983). However, for systems of DDEs, results on balanced realizations have been lacking. Here, the developed Gramians are applied to the problem of the balanced realization for systems of DDEs for the first time. Let $\mathbf{T}$ be a nonsingular state transformation, then $\hat{\mathbf{x}}(t) = \mathbf{T}\mathbf{x}(t)$. The corresponding effect on the Gramians is

$$\hat{C}_o(0, t_1) = \mathbf{T} C_o(0, t_1) \mathbf{T}^T, \quad \hat{O}_b(0, t_1) = \mathbf{T}^{-T} O_b(0, t_1) \mathbf{T}^{-1} \qquad (4.17)$$

Thus, $\hat{C}_o(0, t_1)$ and $\hat{O}_b(0, t_1)$ can be made equal and diagonal with the aid of a suitably chosen matrix $\mathbf{T}$. In the numerical example in (4.15), when $\mathbf{B} = [1\ 0]^T$ and $\mathbf{C} = [1\ 0]$, the $C_o(0, t_1 = 4)$ and $O_b(0, t_1 = 4)$ are respectively

$$C_o(0, t_1) = \begin{bmatrix} 0.2992 & 0.1079 \\ 0.1079 & 0.0554 \end{bmatrix}, \quad O_b(0, t_1) = \begin{bmatrix} 0.2992 & -0.1484 \\ -0.1484 & 0.0975 \end{bmatrix} \quad (4.18)$$

when computed using 11 branches of the matrix Lambert W function. In this case, using the result in (4.17), the transformation

$$\mathbf{T} = \begin{bmatrix} -0.3929 & 1.1910 \\ 1.0880 & -0.5054 \end{bmatrix} \quad (4.19)$$

makes the the Gramians *balanced*, i.e., equal to each other and diagonalized,

$$\hat{C}_o(0, t_1) = \hat{O}_b(0, t_1) = \begin{bmatrix} 0.0238 & 0 \\ 0 & 0.2497 \end{bmatrix} \quad (4.20)$$

Future research is needed to establish conditions for the existence of the transformation $\mathbf{T}$ to achieve a balanced realization for DDEs, and to study its convergence as the number of branches used in the Lambert W function solution increases.

4.5 Conclusions and Future Work

The controllability and observability of linear systems of DDEs is studied using the solution form based on the matrix Lambert W function. The necessary and sufficient conditions for point-wise controllability and observability are derived based on the solution of DDEs (see the summary and comparison to ODEs in Table 4.1). The analytical expressions of Gramians are obtained and approximated for application to real systems with time-delay. Using Gramian concepts, it is possible to figure out how the change in some specific parameters of the system or the delay time, h, affect the controllability and observability of the system via the changes in the Gramians. Also, for the first time, for systems of DDEs, the balanced realization is investigated in the time domain as in the case of ODEs. An example is presented to demonstrate the theoretical results.

Table 4.1 Comparison of the criteria for controllability and observability for the systems of ODEs and DDEs.

ODEs
Controllability
$C_o(0, t_1) \equiv \int_0^{t_1} e^{\mathbf{A}(t_1-\xi)} \mathbf{B}\mathbf{B}^T \left\{ e^{\mathbf{A}(t_1-\xi)} \right\}^T d\xi$
$(s\mathbf{I} - \mathbf{A})^{-1} \mathbf{B}$
$e^{\mathbf{A}(t-0)}\mathbf{B}$
Observability
$O_b(0, t_1) \equiv \int_0^{t_1} \left\{ e^{\mathbf{A}(\xi-0)} \right\}^T \mathbf{C}^T \mathbf{C} e^{\mathbf{A}(\xi-0)} d\zeta$
$\mathbf{C}(s\mathbf{I} - \mathbf{A})^{-1}$
$\mathbf{C}e^{\mathbf{A}(t-0)}$
DDEs
Point-Wise Controllability
$C_o(0, t_1) \equiv \int_0^{t_1} \sum_{k=-\infty}^{\infty} e^{\mathbf{S}_k(t_1-\xi)} \mathbf{C}_k^N \mathbf{B}\mathbf{B}^T \left\{ \sum_{k=-\infty}^{\infty} e^{\mathbf{S}_k(t_1-\xi)} \mathbf{C}_k^N \right\}^T d\xi$
$\left(s\mathbf{I} - \mathbf{A} - \mathbf{A}_d e^{-sh}\right)^{-1} \mathbf{B}$
$\sum_{k=-\infty}^{\infty} e^{\mathbf{S}_k(t-0)} \mathbf{C}_k^N \mathbf{B}$
Point-Wise Observability
$O_b(0, t_1) \equiv \int_0^{t_1} \left\{ \sum_{k=-\infty}^{\infty} e^{\mathbf{S}_k(\xi-0)} \mathbf{C}_k^N \right\}^T \mathbf{C}^T \mathbf{C} \sum_{k=-\infty}^{\infty} e^{\mathbf{S}_k(\xi-0)} \mathbf{C}_k^N d\xi$
$\mathbf{C}\left(s\mathbf{I} - \mathbf{A} - \mathbf{A}_d e^{-sh}\right)^{-1}$
$\mathbf{C} \sum_{k=-\infty}^{\infty} e^{\mathbf{S}_k(t-0)} \mathbf{C}_k^N$

Based upon the results presented, extension of well-established control design concepts for systems of ODEs to systems of DDEs appears feasible. For example, the design of feedback controllers and observers for DDEs can be developed in a manner analogous to ODEs via eigenvalue assignment (Yi *et al.*, 2010b,a) (also see Chapters V and VII).

Chapter 5

Eigenvalue Assignment via the Lambert W Function for Control of Time-Delay Systems

In this chapter, the problem of feedback controller design via eigenvalue assignment for linear time-invariant systems of delay differential equations (DDEs) with a single delay is considered. Unlike ordinary differential equations (ODEs), DDEs have an infinite eigenspectrum and it is not feasible to assign all closed-loop eigenvalues. However, an approach is developed to assign a critical subset of eigenvalues using a solution to linear systems of DDEs in terms of the matrix Lambert W function. The solution has an analytical form expressed in terms of the parameters of the DDE, and is similar to the state transition matrix in linear ODEs. Hence, one can extend controller design methods developed based upon the solution form of systems of ODEs to systems of DDEs, including the design of feedback controllers via eigenvalue assignment. Such an approach is presented here, is illustrated using some examples, and is compared with other existing methods.

5.1 Introduction

Using the classical pole placement method, if a system of linear ordinary differential equations (ODEs) is completely controllable, the eigenvalues can be arbitrarily assigned via state feedback (Chen, 1984). However, delay differential equations (DDEs) always lead to an infinite spectrum of eigenvalues, and the determination of this spectrum requires a corresponding determination of roots of the infinite-dimensional characteristic equation. Moreover, an analytical solution of systems of DDEs has been lacking. Thus, such a pole placement method for controller design for systems of ODEs cannot be applied directly to systems of DDEs.

During recent decades, the stabilization of systems of linear DDEs using feedback control has been studied extensively. The problem of robust stabilization of time-delay systems, or the stabilization problem via delayed feedback control, is most frequently solved via the Finite Spectrum Assignment method (Brethe and Loiseau, 1998; Manitius and Olbrot, 1979;

Wang *et al.*, 1995), which transforms the problem into one for a non-delay system. The stabilization problem can also be approached using stability conditions as expressed by solving a Riccati equation (Lien *et al.*, 1999), or by the feasibility of a set of linear matrix inequalities (Li and deSouza, 1998; Niculescu, 2001). A stability analysis called the *Direct Method*, in which a simplifying substitution is used for the transcendental terms in the characteristic equation (Olgac and Sipahi, 2002), was applied for active vibration suppression by Sipahi and Olgac (2003a). The act-and-wait control concept was introduced for continuous-time control systems with feedback delay by Stepan and Insperger (2006). The study showed that if the duration of waiting is larger than the feedback delay, the system can be represented by a finite dimensional monodromy matrix and, thus, the infinite dimensional pole placement problem is reduced to a finite dimensional one. Also, variants of the Smith predictor method have been developed to decrease errors enabling one to design Proportional-Integral-Derivative (PID) control in time-delay systems (Fliess *et al.*, 2002; Sharifi *et al.*, 2003). A numerical stabilization method was developed by Michiels *et al.* (2002) using a simulation package that computes the rightmost eigenvalues of the characteristic equation. The approach is similar to the classical pole-assignment method for ODEs in determining the rightmost eigenvalues of a linear time-delay system using analytical and numerical methods.

As introduced in Chapter 2, an approach for the solution of linear time invariant systems of DDEs has been developed using the Lambert W function (Asl and Ulsoy, 2003; Yi *et al.*, 2007d). The approach using the Lambert W function provides a solution form for DDEs similar to that of the transition matrix for ODEs (see Table 2.2). Unlike results obtained using other existing methods, the solution has an analytical form expressed in terms of the parameters of the DDE. One can determine how the parameters are involved in the solution and, furthermore, how each parameter affects each eigenvalue and the solution. Also, each eigenvalue in the infinite eigenspectrum is associated with a branch of the Lambert W function. Hence, the concept of the state transition matrix in ODEs can be generalized to DDEs using the matrix Lambert W function. This suggests that some analyses used in systems of ODEs, based upon the concept of the state transition matrix, can potentially be extended to systems of DDEs.

In this chapter, the matrix Lambert W function-based approach for the solution to DDEs is applied to stabilize linear systems of DDEs. A new approach for controller design via eigenvalue assignment of systems of DDEs is presented, and the method is illustrated with several examples.

Using the proposed method, it is possible to move a dominant subset of the eigenvalues to desired locations in a manner similar to pole placement for systems of ODEs. For a given system represented by DDEs, the solution to the system is obtained based on the Lambert W function, and stability is determined. If the system is unstable, after controllability of the system is checked, a stabilizing feedback is designed by assigning eigenvalues, and finally the closed-loop system of DDEs can be stabilized. These processes are conducted based upon the Lambert W function-based approach.

5.2 Eigenvalue Assignment for Time-Delay Systems

5.2.1 *Stability*

Consider a linear time invariant (LTI) real system of delay differential equations with a single constant delay, h, in Eq. (2.39). The *Conjecture* in Subsection 3.3.1 has been observed consistently in all the examples that have been considered. The *Conjecture* was formulated as the basis not only to determine the stability of systems of DDEs, but also to place a subset of the eigenspectrum at desired locations as presented in this chapter.

A major difficulty with designing a feedback controller for a time-delay system is assigning all of the eigenvalues. This difficulty is due to the infinite spectrum of eigenvalues and a finite number of control paramteters (Manitius and Olbrot, 1979). Placing a selected finite number of eigenvalues by classical pole placement method for ODEs (Chen, 1984) may cause other uncontrolled eigenvalues to move to the right half plane (RHP) (Michiels *et al.*, 2002). However, the approach presented for control design using the matrix Lambert W function, based on the *Conjecture* in Subsection 3.3.1, provides proper control laws without such loss of stability.

5.2.2 *Eigenvalue assignment*

First, consider a free first-order scalar DDE, as in Eq. (1.2). The solution to Eq. (1.2) can be obtained using the Lambert W function as in Eq. (1.3). And the roots, λ_k, of the characteristic equation of Eq. (1.2), $\lambda - a - a_d e^{-\lambda h} = 0$, are given by

$$\lambda_k = \frac{1}{h} W_k(a_d h e^{-ah}) + a, \quad \text{for } k = -\infty, \cdots, -1, 0, 1, \cdots, \infty \quad (5.1)$$

This solution is exact and analytical. In this scalar case, one can compute all of the roots and the rightmost pole among them is always obtained by using the principal branch $(k = 0)$, and the pole determines stability of Eq. (1.2) (Shinozaki and Mori, 2006), that is,

$$\max\left[\mathrm{Re}\{W_k(H)\}\right] = \mathrm{Re}\{W_0(H)\} \tag{5.2}$$

In designing a control law for delayed system, it is crucial to handle the rightmost poles among an infinite set. In this regard, the property in (5.2) of the Lambert W function provides a useful basis for assigning the rightmost pole. By adjusting the parameters, a, a_d and/or the delay time, h, one can assign the rightmost pole of the system to the desired values in the complex plane based on Eq. (5.2). First, decide on the desired location of the rightmost pole, λ_{des}, thus equate it to the pole corresponding to $k = 0$, that is,

$$\lambda_{des} = \lambda_0 = \frac{1}{h}W_0(a_d h e^{-ah}) + a \tag{5.3}$$

Equation (5.3) can be solved using numerical methods, e.g., commands already embedded in Matlab, such as *fsolve* and *lambertw*.

Example 5.1. Consider the scalar DDE in Eq. (1.2) with $a = -1$ and $h = 1$. Table 5.1 shows the corresponding values of a_d required to move the rightmost pole of equation to the exact desired locations. As seen in Fig. 5.1, each rightmost pole is located at the desired position corresponding to a value of a_d.

However, each branch of the Lambert W function has its own range and, especially, the value of the principal branch has the range (e.g., see Fig. 1.1):

$$\mathrm{Re}\{W_0(H)\} \geq -1 \tag{5.4}$$

Therefore, depending on the structure or parameters of a given system, there exist limitations on assigning the rightmost pole. Although general research on the limitation is lacking so far (see also Chapter 7 and Appendix C), in the example above it can be concluded that using Eq. (5.4)

$$\mathrm{Re}\{S_0\} = \frac{1}{h}\underbrace{\mathrm{Re}\left\{W_0(a_d h e^{-ah})\right\}}_{\geq -1} + a \geq -\frac{1}{h} + a \tag{5.5}$$

Thus, the rightmost eigenvalue cannot be smaller than -2 for any value of a_d.

In the case of systems of DDEs, by adjusting the elements in the coefficient matrices in Eq. (2.39), one can assign the eigenvalues of a single

Table 5.1   Corresponding values of a_d for each desired pole. By adjusting the parameter, a_d, it is possible to assign the rightmost pole of the system in Eq. (1.2) to the desired values.

λ_{des}	-0.5	0	0.5
a_d	0.3033	1.0000	2.4731

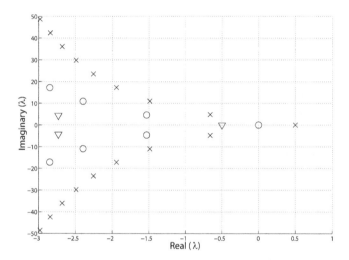

Fig. 5.1   Eigenspectra of Eq. (1.2) with $a = -1$, $h = 1$ and a_d, in Table 5.1. Using the values of a_d, it is possible make the rightmost poles move exactly to the desired locations $(-0.5(\triangledown); 0 \, (O); 0.5(X))$.

matrix corresponding to the principal branch, $\mathbf{S}_0$, by solving simultaneously Eqs. (2.19) and (2.20), and

$$\text{Eigenvalues of } \mathbf{S}_0 = \text{desired values} \tag{5.6}$$

using numerical methods embedded in software packages, such as *fsolve* in Matlab. This approach is based upon the *Conjecture* presented previously in Chapter 3, and applied here to design feedback controllers and is validated with examples. In the subsequent section, the approach presented above is applied to the design the feedback control laws for both scalar DDEs and systems of DDEs.

5.3   Design of a Feedback Controller

Delay terms can arise in two different ways: i) delays in the control, that is, $\mathbf{u}(t - h)$, or ii) delays in the state variables, $\mathbf{x}(t - h)$. In both cases the resulting feedback operators contain integrals over the past values of control or state trajectory (Manitius and Olbrot, 1979). For systems without time delay, in addition, a time-delayed control is often used for various special purposes, often motivated by intuition. The most common examples are the vibration absorber with delayed feedback control, with which is possible to absorb an external force of unknown frequencies (Olgac *et al.*, 1997), and delayed feedback to stabilize unstable periodic orbits without any information of the periodic trajectory except the period by constructing a control force from the difference of the current state to the state one period before (Hovel and Scholl, 2005; Pyragas, 1992). For those systems, the approach for eigenvalue assignment for time-delay systems can be important.

5.3.1   *Scalar case*

For a scalar DDE with state feedback

$$\begin{aligned} \dot{x}(t) &= ax(t) + a_d x(t - h) + u(t) \\ u &= kx(t) \end{aligned} \tag{5.7}$$

One may try to design the feedback control, $u = kx(t)$, by using the first-order Padé approximation as

$$e^{-hs} \approx \frac{1 - hs/2}{1 + hs/2} \tag{5.8}$$

Then, the characteristic equation of (5.7) becomes a simple 2nd order polynomial as

$$s^2 h + s(2 - ah - kh + a_d h) - 2(a + k) - 2a_d = 0 \tag{5.9}$$

Then, for a desired pole, one can obtain the control gain. For example, with parameters $a = 1$, $a_d = -2$, and $h = 1$, Eq. (5.9) becomes

$$s^2 - s(1 + k) + 2 - 2k = 0 \tag{5.10}$$

For the value $k = -1.1$, Eq. (5.10) has two stable poles. However, this control gain is applied to the original system (5.7) will fail to stabilize the system. The resulting eigenspectrum is shown in Fig. 5.2. Even though higher order Padé approximations, or other advanced rational approximations, can be used to approximate the exponential term in the characteristic

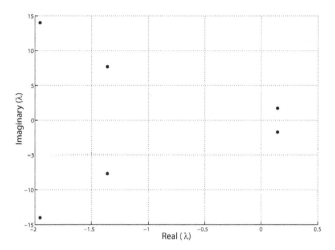

Fig. 5.2 Eigenspectrum with the feedback controller designed with Padé approximation shows the failure in moving the pole to the desired value and thus stabilizing the system.

equation more precisely, such approaches are limited by an inevitable limitation in accuracy, and at worst may lead to instability of the original system (Richard, 2003; Silva and Datta, 2001).

Alternatively, from the characteristic equation of (5.7) with the desired pole, the linear equation for k can be derived as

$$\lambda_{des} - a - a_d e^{-\lambda_{des}h} - k = 0 \tag{5.11}$$

For example, with the parameters $a = 1$, $a_d = -1$, and $h = 1$, just by substituting the variable as $\lambda_{des} = -1$, then, the obtained gain, k, using Eq. (5.11) is 0.7183. However, this control gain is also applied to the original system in (5.7) and fails to stabilize the system, because the desired pole is not guaranteed to be the rightmost pole. The resulting eigenspectrum is shown in Fig. 5.3. While one of the poles, not rightmost, is placed at the desired location, -1, the rightmost one is in the RHP. Although for the desired pole, the control gain, k, is derived, when the gain is applied to (5.11), there exists other infinite number of poles to satisfy the equation, some of them can have real parts larger than that of the desired pole.

On the other hand, using the Lambert W function one can safely assign the real part of the rightmost pole exactly. For example, for the system (5.7) with $a = 1$, $a_d = -1$, and $h = 1$,

$$\text{Re}(S_0 = \frac{1}{h}W_0(a_d h e^{-(a+k)h}) + a + k) = -1 \tag{5.12}$$

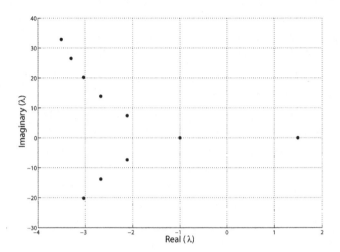

Fig. 5.3   Eigenspectrum with the feedback controller designed with the linear approach in (5.11) shows the failure in moving the pole to the desired value and thus stabilizing the system.

Then, the resulting value of k is -3.5978. As seen in Fig. 5.4, the rightmost eigenvalues are placed at the exact desired location. Compared with the results in Fig. 5.2 and Fig. 5.3, the approach using the Lambert W function provides the exact result and stabilizes the unstable system safely. In the next two subsections, this approach is generalized to two different types of systems of DDEs.

5.3.2   *Systems with control delays*

In systems of controllable ODEs, one of the significant results of control theory is that, with full state feedback, one can specify all the closed-loop eigenvalues arbitrarily by selecting the gains. However, systems of DDEs have an infinite number of eigenvalues and it is not feasible to specify all of them with linear controllers, which have a finite number of gains. Furthermore, research on the explicit relation between controllability and eigenvalue assignment is lacking so far (refer to Chapter 4). Nevertheless, in this subsection and the next, for the controllable system of DDEs, the Lambert W function-based approach is used to specify the first matrix, $\mathbf{S}_0$, corresponding to the principal branch, $k = 0$, and observed to be critical in the solution form in Eq. (2.43), by choosing the feedback gain and designing a feedback controller.

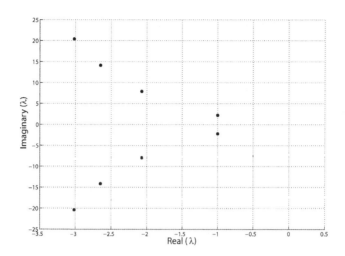

Fig. 5.4   Eigenspectrum with the feedback controller designed using the Lambert W function approach. The rightmost eigenvalues are placed at the exact desired location.

First, consider the system of DDEs:

$$\dot{\mathbf{x}}(t) = \mathbf{A}\mathbf{x}(t) + \mathbf{B}\mathbf{u}(t - h) \tag{5.13}$$

Then, the feedback,

$$\mathbf{u}(t) = \mathbf{K}\mathbf{x}(t) \tag{5.14}$$

yields the a closed-loop form

$$\dot{\mathbf{x}}(t) = \mathbf{A}\mathbf{x}(t) + \mathbf{B}\mathbf{K}\mathbf{x}(t - h) \tag{5.15}$$

The gain, $\mathbf{K}$, to assign the rightmost eigenvalues is determined as follows. First, select desired eigenvalues, $\lambda_{i,des}$ for $i = 1, , n$, and set an equation so that the selected eigenvalues become those of the matrix $\mathbf{S}_0$ as

$$\lambda_i(\mathbf{S}_0) = \lambda_{i,des}, \quad \text{for} \quad i = 1, \ldots, n \tag{5.16}$$

where, $\lambda_i(\mathbf{S}_0)$ is i^{th} eigenvalue of the matrix $\mathbf{S}_0$. Second, apply the two new coefficient matrices $\mathbf{A}' \equiv \mathbf{A}$, $\mathbf{A}'_{\mathbf{d}} \equiv \mathbf{B}\mathbf{K}$ in Eq. (5.15) to Eq. (2.19) and solve numerically to obtain the matrix $\mathbf{Q}_0$ for the principal branch $(k = 0)$. Note that $\mathbf{K}$ is an unknown matrix with all unknown elements in it, and the matrix $\mathbf{Q}_0$ is a function of the unknown $\mathbf{K}$. Then, for the third step, substitute the matrix $\mathbf{Q}_0$ from Eq. (2.19) into Eq. (2.20), to obtain $\mathbf{S}_0$ and its eigenvalues as the function of the unknown matrix $\mathbf{K}$. Finally, Eq. (5.16) with the matrix, $\mathbf{S}_0$, is solved for the unknown $\mathbf{K}$ using numerical methods, such as *fsolve* in Matlab. As mentioned previously in this chapter, depending on the structure or parameters of given system, there exists a limitation of the rightmost eigenvalues and some values are not proper for the rightmost eigenvalues. In that case, the above approach does not yield any solution for **K**. To resolve the problem, one may try again with fewer desired eigenvalues, or different values of the desired rightmost eigenvalues. Then, the solution, **K**, is obtained numerically for a variety of initial conditions by an iterative trial and error procedure.

Example 5.2. Consider the van der Pol equation, which has become a prototype for systems with self-excited limit cycle oscillations and has the form of

$$\ddot{x}(t) + f(x,t)\dot{x}(t) + x(t) = g(x,t;h) \tag{5.17}$$

with

$$f(x,t) = \varepsilon \left(x^2(t) - 1 \right) \tag{5.18}$$

For the dynamics of the van der Pol equation under the effect of linear position and velocity time delayed feedback, the left side of Eq. (5.17) can be written as

$$g(x,t;h) = k_1 x(t-h) + k_2 \dot{x}(t-h) \tag{5.19}$$

Then, with the damping coefficient function in (5.18) and feedback in (5.19), Eq. (5.17) becomes

$$\ddot{x}(t) + x(t) = \varepsilon \left(1 - x^2(t) \right) \dot{x}(t) + k_1 x(t-h) + k_2 \dot{x}(t-h) \tag{5.20}$$

Linearizing Eq. (5.20) about the zero equilibrium yields the equation for infinitesimal perturbations,

$$\ddot{x}(t) + x(t) = \varepsilon \dot{x}(t) + k_1 x(t-h) + k_2 \dot{x}(t-h) \tag{5.21}$$

Or, equivalently, by defining $x_1 = x$ and $x_2 = \dot{x}$, one obtains the state equations

$$\left\{ \begin{matrix} \dot{x}_1(t) \\ \dot{x}_2(t) \end{matrix} \right\} = \begin{bmatrix} 0 & 1 \\ -1 & \varepsilon \end{bmatrix} \left\{ \begin{matrix} x_1(t) \\ x_2(t) \end{matrix} \right\} + \begin{bmatrix} 0 & 0 \\ k_1 & k_2 \end{bmatrix} \left\{ \begin{matrix} x_1(t-h) \\ x_2(t-h) \end{matrix} \right\} \tag{5.22}$$

which can also be expressed in the form of (5.15) as

$$\dot{\mathbf{x}}(t) = \underbrace{\begin{bmatrix} 0 & 1 \\ -1 & \varepsilon \end{bmatrix}}_{\mathbf{A}} \mathbf{x}(t) + \underbrace{\begin{bmatrix} 0 \\ 1 \end{bmatrix}}_{\mathbf{B}} \underbrace{\begin{bmatrix} k_1 & k_2 \end{bmatrix}}_{\mathbf{K}} \mathbf{x}(t-h) \tag{5.23}$$

Equations of this type have been investigated using the asymptotic perturbation method (Maccari, 2001), bifurcation methods (Reddy *et al.*, 2000; Wirkus and Rand, 2002; Xu and Chung, 2003) and a Taylor expansion with averaging (Li *et al.*, 2006) to show that vibration control and quasi-periodic motion suppression are possible for appropriate choices of the time delay and feedback gains. The effect of time-delays under an external excitation with various practical examples was considered by (Maccari, 2003), demonstrating the importance of this oscillator in engineering science.

Controllability and stabilizability of the system of equations in (5.13) have been studied during recent decades (Frost, 1982; Mounier, 1998; Olbrot, 1972). For example, the linear system of (5.13) is said to be controllable on $[0, t_1]$ if there exists an *admissible* (that is, measurable and bounded on a finite time interval) control $\mathbf{u}(t)$ such that $\mathbf{x}(t_1) = 0$ where $t_1 > h$. According to the criterion presented by Olbrot (1972), the system is controllable on $[0, t_1]$ if and only if

$$\text{rank}\,[\mathbf{B} \,\vdots\, \mathbf{AB}] = n \tag{5.24}$$

According to the definition and the corresponding simple rank condition in Eq. (5.24), the system of Eq. (5.22) is controllable. Thus, using the pole placement method introduced in the previous section, one can design an appropriate feedback controller to stabilize the system and choose the gains, k_1 and k_2, to locate the eigenvalues at desired positions in the complex plane.

Without the delayed feedback term (i.e., $k_1 = k_2 = 0$), the system in (5.22) is unstable when $\varepsilon = 0.1$, and its rightmost eigenvalues are $0.0500 \pm 0.9987i$. For example, when $h = 0.2$ if the desired eigenvalues are -1 and -2, which are arbitrarily selected, then, the required gains are $k_1 = -0.0469$, $k_2 = -1.7663$ found by using the presented Lambert W function-based approach. As seen in Fig. 5.5, the response without feedback control is unstable. Applying the designed feedback controller stabilizes the system. Figure 5.6 shows the eigenspectra of systems without feedback and with feedback. The rightmost eigenvalues are moved exactly to the desired locations and all the other eigenvalues are to the left. If the desired eigenvalues are $-1 \pm 2i$, or $-1 \pm 1i$, then the corresponding gains are $\mathbf{K} = [-1.9802 \quad -1.8864]$, or, $\mathbf{K} = [-0.2869 \quad -1.5061]$, respectively.

In (Michiels *et al.*, 2002), a numerical stabilization method was developed using a simulation package that computes the rightmost eigenvalues of the characteristic equation. For the obtained finite number of eigenvalues, the eigenvalues can be moved to the LHP using sensitivities with respect to

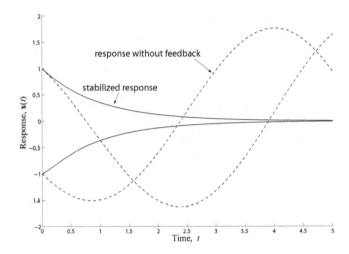

Fig. 5.5   Comparison of responses before (dashed) and after (solid) applying feedback in Eq. (5.19) with $\mathbf{K} = [-0.0469 \quad -1.7663]$. Chosen feedback gain stabilizes the system.

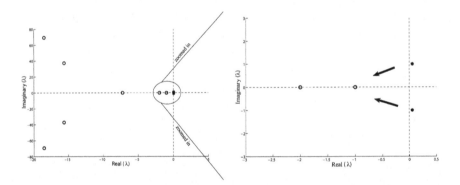

Fig. 5.6   Movement of eigenvalues after applying the feedback ($*$ without feedback; $\circ$ with feedback). The rightmost eigenvalues are located at the exact desired location -1 and -2.

changes in the feedback gain, $\mathbf{K}$ (see Figs. 5.7 and 5.8). Compared with the approach, the matrix Lambert W function-based method yields the equation for assignment of the rightmost eigenvalues with the parameters of the system. Using the analytical expression, one can obtain the control gain to move the critical eigenvalues to the desired positions without starting with their initial unstable positions or computing the rightmost eigenvalues

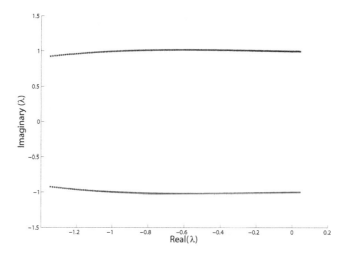

Fig. 5.7   Movement of eigenvalues from their original positions, $0.0500 \pm 0.9987i$, to LHP when the method in (Michiels *et al.*, 2002), is applied to the system in (5.23).

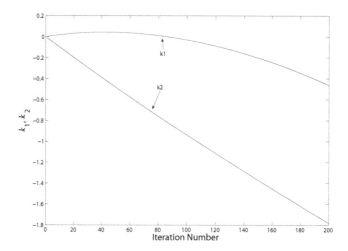

Fig. 5.8   The value of **K** in Eq. (5.23) corresponding to the number of iteration to compute for the movement of eigenvalues in Fig. 5.7.

and their sensitivities after every small movement in a quasi-continuous way. Using the Lambert W function, one can find the control gain independently of the path of the rightmost eigenvalues. Without planning the path, only from the destination, the control laws for the system are obtained.

For the system in Eq. (5.13) with Eq. (5.14), the control law by Finite Spectrum Assignment (FSA) method based on prediction, which is given by

$$\mathbf{u}(t) = \mathbf{K}e^{\mathbf{A}h}\mathbf{x}(t) + \mathbf{K}\int_0^h e^{\mathbf{A}(h-\theta)}\mathbf{B}\mathbf{u}(t+\theta-h)d\theta \qquad (5.25)$$

can make the system finite dimensional and the assign the finite eigenvalues to be desired values (Brethe and Loiseau, 1998; Manitius and Olbrot, 1979; Wang *et al.*, 1995). However, such a method requires model-based calculation, which may cause unexpected errors when applied to a real system. Limitations on FSA have been studied with several examples in (Engelborghs *et al.*, 2001) and (Van Assche *et al.*, 1999), the implementation of such controller is still an open problem (Richard, 2003).

5.3.3   *Systems with state delays*

Consider the following time delayed system:

$$\dot{\mathbf{x}}(t) = \mathbf{A}\mathbf{x}(t) + \mathbf{A_d}\mathbf{x}(t-h) + \mathbf{B}\mathbf{u}(t) \qquad (5.26)$$

and a generalized feedback containing current and delayed states:

$$\mathbf{u}(t) = \mathbf{K}\mathbf{x}(t) + \mathbf{K_d}\mathbf{x}(t-h) \qquad (5.27)$$

Then, the closed-loop system becomes

$$\dot{\mathbf{x}}(t) = (\mathbf{A}+\mathbf{B}\mathbf{K})\mathbf{x}(t) + (\mathbf{A_d}+\mathbf{B}\mathbf{K_d})\mathbf{x}(t-h) \qquad (5.28)$$

The gains, $\mathbf{K}$ and $\mathbf{K_d}$ are determined in a way similar to the previous subsection as follows. First, select desired eigenvalues, $\lambda_{i,des}$ for $i = 1, \ldots, n$, and set an equation so that the selected eigenvalues become those of the matrix $\mathbf{S}_0$ as

$$\lambda_i(\mathbf{S}_0) = \lambda_{i,des}, \quad \text{for} \quad i = 1, \ldots, n \qquad (5.29)$$

where, $\lambda_i(\mathbf{S}_0)$ is i^{th} eigenvalue of the matrix $\mathbf{S}_0$. Second apply the new two coefficient matrices $\mathbf{A}' + \mathbf{B}\mathbf{K}$ and $\mathbf{A}'_\mathbf{d} + \mathbf{B}\mathbf{K_d}$ in Eq. (5.28) to Eq. (2.19) and solve numerically to obtain the matrix $\mathbf{Q}_0$ for the principal branch ($k = 0$). Note that $\mathbf{K}$ and $\mathbf{K_d}$ are unknown matrices with all unknown elements, and the matrix $\mathbf{Q}_0$ is a function of the unknown $\mathbf{K}$ and $\mathbf{K_d}$. For the third step, substitute the matrix $\mathbf{Q}_0$ from Eq. (2.19) into Eq. (2.20) to obtain $\mathbf{S}_0$ and its eigenvalues as the function of the unknown matrix $\mathbf{K}$ and $\mathbf{K_d}$. Finally, Eq. (5.29) with the matrix, $\mathbf{S}_0$, is solved for the unknown $\mathbf{K}$ and $\mathbf{K_d}$ using numerical methods, such as *fsolve* in Matlab. As mentioned in Section 5.2, depending on the structure or parameters of given system, there exist limitations on the location of the rightmost eigenvalues, and some values are not proper for the rightmost eigenvalues. In that case, the above approach does not yield any solution for $\mathbf{K}$ and $\mathbf{K_d}$. To resolve the problem, one may try again with fewer desired eigenvalues, or different values of the desired rightmost eigenvalues. Then, the solution, $\mathbf{K}$ and $\mathbf{K_d}$, is obtained numerically for a variety of initial conditions by an empirical trial and error procedure.

The controllability of such a system, using the solution form of Eq. (2.39), was studied by Yi *et al.* (2008a) as presented in Chapter 4. In the case of LTI systems of ODEs, if it is completely controllable, then the eigenvalues can arbitrarily be assigned by choosing feedback gain. Here examples regarding controllability and eigenvalue assignment in DDEs are considered.

Example 5.3. Consider the following system of DDEs,

$$\dot{\mathbf{x}}(t) = \begin{bmatrix} 1.1 & -0.1732 \\ -0.0577 & 1.1 \end{bmatrix} \mathbf{x}(t) + \begin{bmatrix} 0.35 & 0.2598 \\ 0.0866 & 0.35 \end{bmatrix} \mathbf{x}(t-h) + \begin{bmatrix} 1 \\ -0.5774 \end{bmatrix} u(t)$$

(5.30)

When the coefficients are applied to the condition in (4.11) for controllability, the corresponding matrix is

$$(s\mathbf{I} - \mathbf{A} - \mathbf{A_d}e^{-sh})^{-1}\mathbf{B} = \begin{bmatrix} \dfrac{5}{5s - 6 - e^{-sh}} \\ \dfrac{-5/\sqrt{3}}{5s - 6 - e^{-sh}} \end{bmatrix}$$

(5.31)

Obviously, the two elements in Eq. (5.31) are linearly dependent and Eq. (5.30) fails the rank condition in (4.11). Thus, the system in Eq. (5.30) is not point-wise controllable and one cannot find any appropriate feedback control in the form of (5.27) to stabilize it.

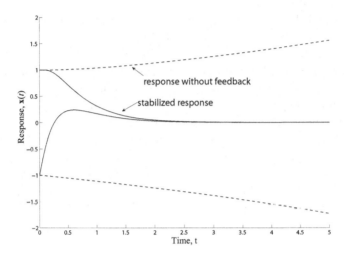

Fig. 5.9   Comparison of responses before (dashed) and after (solid) applying feedback (5.27) with feedback gains $\mathbf{K} = [\,-0.1687 \quad -3.6111]$ and $\mathbf{K_d} = [1.6231 \quad -0.9291]$. The chosen feedback gain stabilizes the system.

Example 5.4. Consider the following time-delay model, from (Mahmoud and Ismail, 2005)

$$\dot{\mathbf{x}}(t) = \begin{bmatrix} 0 & 0 \\ 0 & 1 \end{bmatrix} \mathbf{x}(t) + \begin{bmatrix} -1 & -1 \\ 0 & -0.9 \end{bmatrix} \mathbf{x}(t-h) + \begin{bmatrix} 0 \\ 1 \end{bmatrix} \mathbf{u}(t) \qquad (5.32)$$

Before applying the feedback, the two rightmost eigenvalues are -1.1183 and 0.1098, and, thus, the system is unstable when the delay time, $h = 0.1$ (see Figs. 5.9 and 5.10). When the coefficients are applied to the condition in (4.11), the system in Eq. (5.32) satisfies the criterion, and, thus, is pointwise controllable. Then, using the pole placement method, one can design an appropriate feedback controller to stabilize the system and choose the gains $\mathbf{K}$ and $\mathbf{K_d}$ to locate the eigenvalues at desired positions in the complex plane. For example, when the desired eigenvalues are -1 and -6, which are chosen arbitrarily, the gains obtained by using the presented approach are $\mathbf{K} = [\,-0.1391 - 1.8982]$; $\mathbf{K_d} = [\,-0.1236 - 1.8128]$, or $\mathbf{K} = [\,-0.1687 - 3.6111]$; $\mathbf{K_d} = [1.6231 - 0.9291]$ for -2 and -4. By applying the obtained feedback gains to Eq. (5.27), one can stabilize the system (see Fig. 5.9) and place the eigenvalues at a desired positions in the complex plane (see Fig. 5.10).

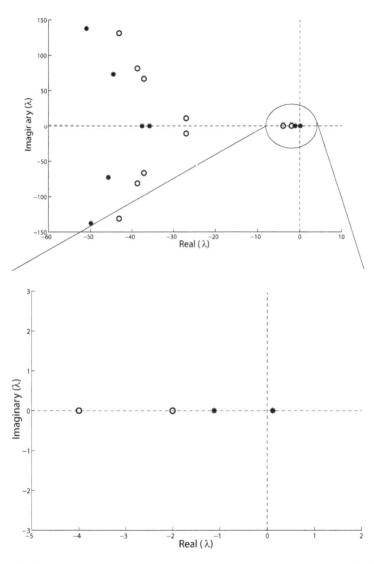

Fig. 5.10 Movement of eigenvalues after applying feedback (∗ without feedback; ○ with feedback). The rightmost eigenvalues are located at the exact desired location, −2.0000 and −4.0000, using the feedback gains, $\mathbf{K} = [-0.1687 \quad -3.6111]$ and $\mathbf{K_d} = [1.6231 \quad -0.9291]$.

5.4   Conclusions

In this chapter, new results for feedback controller design for a class of time-delay systems are presented. For a given system, which can be represented by DDEs, based on the Lambert W function, the solution to the system is obtained, and stability is subsequently determined. If the system is unstable, after the controllability of the system is checked, a stabilizing feedback is designed by assigning eigenvalues, and finally the closed-loop system of DDEs can be stabilized. All of these results are based upon the Lambert W function-based approach. Numerical examples are presented to illustrate the approach. Although DDEs have an infinite eigenspectrum, and it is not possible to assign all closed-loop eigenvalues, it is possible to assign a subset of them, i.e., the rightmost or dominant eigenvalues critical for determining stability and performance.

The proposed method, based upon the Lambert W function, is compared with other approaches (see examples in Section 5.3). Many of these are ad-hoc, and can fail on certain problems (Richard, 2003; Silva and Datta, 2001). The FSA method is based upon prediction, and known to have implementation problems (Engelborghs *et al.*, 2001; Van Assche *et al.*, 1999). The method of Michiels *et al.* (2002) is the most effective, but is an iterative method based upon sensitivity of eigenvalues to the control gains. The Lambert W function-based method is direct and effective in all problems evaluated.

The presented approach is extended for the design of systems with observer-based feedback controller for systems of DDEs, and problems of robust controller design and time-domain specifications in the following chapters.

Chapter 6

Robust Control and Time-Domain Specifications for Systems of Delay Differential Equations via Eigenvalue Assignment

An approach to eigenvalue assignment for systems of delay differential equations (DDEs), based upon the solution in terms of the matrix Lambert W function, is applied to the problem of robust control design for perturbed systems of DDEs, and to the problem of time-domain response specifications. Robust stability of the closed-loop system can be achieved through eigenvalue assignment combined with the real stability radius concept. For a linear system of DDEs with a single delay, which has infinite number of eigenvalues, the recently developed Lambert W function-based approach is used to assign a dominant subset of them, which has not been previously feasible. Also, an approach to time-domain specifications for the transient response of systems of DDEs is developed in a way similar to systems of ordinary differential equations using the Lambert W function-based approach.

6.1 Introduction

A primary goal for control engineers is to maintain the stability of a system, an essential requirement, while achieving good performance to meet response specifications (Suh and Yang, 2005). For a system of delay differential equations (DDEs), even though more complex than systems of ordinary differential equations (ODEs) due to its transcendental characteristic equation, various methods to achieve that goal have been introduced in the literature in recent decades. For a detailed survey, refer to (Yi *et al.*, 2010b; Richard, 2003) and the references therein.

However, systems frequently have uncertainties in model parameters caused by estimation errors, modeling errors, or linearization. For such perturbed systems, it is naturally required to design controllers to make sure that the controlled system remains stable in the presence of such uncertainties.

Usually, the robust control problem for systems of DDEs has been handled by using Lypunov functions, and employing linear matrix inequalities (LMIs) or algebraic Riccati equations (AREs) (see, e.g., (Mahmoud, 2000; Niculescu, 1998) and the references therein). Even though such approaches can be applied to quite general types of time-delay systems (e.g., systems with multiple delays, time-varying delay), they provide only sufficient conditions and are substantially conservative because of their dependence on the selection of cost functions and their coefficients (Michiels and Roose, 2003). Moreover, general systematic procedures to construct appropriate Lyapunov functions are not available, and solving the resulting LMI/ARE can be nontrivial (Hrissagis and Kosmidou, 1998). Analysis and design of control systems in the frequency domain is well established in control engineering. Stability is investigated based on the transfer function and the Nyquist criteria. By computing the robust stability margin in the Nyquist plane, the method has been used for robust control of systems of ODEs (Postlethwaite and Foo, 1985) and, also, DDEs (Wang and Hu, 2007) with uncertainties. However, although being improved extensively, typically the method requires an exhaustive numerical search in the frequency domain plus an exhaustive search in the parameter domain.

On the other hand, while pursuing robust control, one may have to retain the positions of the eigenvalues to meet the response specifications, such as time-domain specifications (see Section 6.3), of the nominal system. For this reason, robust stability of the closed-loop system has often been achieved through eigenvalue assignment combined with robust stability indices (e.g., the stability radius concept in Section 6.2). Such indices set the upper limit on parameter perturbations and help select the positions of the desired eigenvalues in the complex plane. Then using an eigenvalue assignment method, it is possible to find feedback control for robust stabilization for systems of ODEs (e.g., see (Kawabata and Mori, 2009) and the references therein). However, systems of DDEs have an infinite number of eigenvalues, which are the roots of a transcendental equation, and it is not practically feasible to assign all of them. Thus, the usual pole placement design techniques for ODEs cannot be applied without considerable modification to systems of DDEs (Tsoi and Gregson, 1978).

In this chapter, a new approach to design robust controllers for a system of DDEs through eigenvalue assignment based on the Lambert W function approach is presented. An eigenvalue assignment method for systems of DDEs was developed in Chapter 5. Using that approach, one can design a linear feedback controller to place the rightmost eigenvalues at the desired positions in the complex plane and, thus, stabilize systems with a single time-delay. In that study, the critical rightmost subset of eigenvalues, which determine stability of the system, among the infinite eigenspectrum is assigned. This is possible because the eigenvalues are expressed in terms of the parameters of the system and each one is distinguished by a branch of the Lambert W function.

In this chapter, the Lambert W function-based approach to eigenvalue assignment for DDEs, is combined with the stability radius concept to address the problem of robust stability of systems with uncertain parameters (Section 6.2). Also, an approach for improvement of the transient response for systems of DDEs is presented. The method developed in Chapter 5 also makes it possible to assign simultaneously the real and imaginary parts of a critical subset (based on the concept of *dominant poles* (Franklin *et al.*, 2005)) of the eigenspectrum with linear feedback control. Therefore, guidelines, similar to those used for systems of ODEs to improve transient response, can be used for systems of DDEs via eigenvalue assignment by using the matrix Lambert W function-based approach to meet time-domain response specifications (Section 6.3).

6.2 Robust Feedback

6.2.1 *Stability radius*

To design robust feedback controllers through eigenvalue assignment, it is required to decide where is the appropriate positions in the complex plane to guarantee robust stability depending on the size of uncertainty. The decision can be made by using robust stability indices. The real stability radius, which is one of the indices and the norm of the minimum destabilizing perturbations, was obtained for linear systems of ODEs and a computable formula for the exact real stability radius was presented by [Qiu *et al.* (1995)]. The real stability radius measures the ability of a system to preserve its stability under a certain class of real perturbations. The formula was extended to perturbed linear systems of DDEs in (Hu and Davison, 2003).

Assume that the perturbed system (2.1) can be written in the form

$$\begin{aligned}
\dot{\mathbf{x}}(t) &= \{\mathbf{A} + \delta\mathbf{A}\}\mathbf{x}(t) + \{\mathbf{A_d} + \delta\mathbf{A_d}\}\mathbf{x}(t - h) \\
&= \{\mathbf{A} + \mathbf{E}\Delta_1\mathbf{F}_1\}\mathbf{x}(t) + \{\mathbf{A_d} + \mathbf{E}\Delta_2\mathbf{F}_2\}\mathbf{x}(t - h)
\end{aligned} \tag{6.1}$$

where $\mathbf{E} \in R^{n \times m}$, $\mathbf{F}_i \in R^{l_i \times n}$, and $\Delta_i \in R^{m \times l_i}$ denotes the perturbation matrix. Provided that the unperturbed system (2.1) is stable, the real structured stability radius of (6.1) is defined as (Hu and Davison, 2003)

$$r_{\mathbb{R}} = \inf\{\sigma_1(\Delta) : \text{system (6.1) is unstable}\} \tag{6.2}$$

where $\Delta = [\Delta_1 \ \Delta_2]$ and $\sigma_1(\Delta)$ denotes the largest singular value of Δ. The largest singular value, $\sigma_1(\Delta)$ is equal to the operator norm of Δ, which measures the size of Δ by how much it lengthens vectors in the worst case. Thus, the stability radius in Eq. (6.2) represents the size of the smallest perturbations in parameters, which can cause instability of a system. And the real stability radius problem concerns the computation of the real stability radius when the nominal system is known. The stability radius is computed from (Hu and Davison, 2003)

$$r_{\mathbb{R}} = \left\{ \sup_{\omega} \inf_{\gamma \in (0,1]} \sigma_2 \left(\begin{bmatrix} \Re(\Omega(j\omega)) & -\gamma\Im(\Omega(j\omega)) \\ \gamma^{-1}\Im(\Omega(j\omega)) & \Re(\Omega(j\omega)) \end{bmatrix} \right) \right\} \tag{6.3}$$

where

$$\Omega(s) = \begin{bmatrix} \mathbf{F}_1 \\ \mathbf{F}_2 e^{-hs} \end{bmatrix} (s\mathbf{I} - \mathbf{A} - \mathbf{A_d})^{-1}\mathbf{E} \tag{6.4}$$

In (6.3), it is not practically feasible to compute the supremum value for the whole range of $\omega \in (-\infty, \infty)$. However, for the value ω^*, which satisfies

$$\omega^* < \bar{\sigma}(\mathbf{A}) + \bar{\sigma}(\mathbf{A_d}) + \bar{\sigma}(\mathbf{E})\underline{\sigma}[\mathbf{W}(0)]\bar{\sigma}([\mathbf{F}_1\mathbf{F}_2]) \tag{6.5}$$

where $\bar{\sigma}(\cdot)$ and $\underline{\sigma}(\cdot)$ are the largest and smallest singular values of $(\cdot)$, respectively, and

$$\mathbf{W}(0) = \begin{bmatrix} \mathbf{F}_1 \\ \mathbf{F}_2 \end{bmatrix} (-\mathbf{A} - \mathbf{A_d})^{-1}\mathbf{E} \tag{6.6}$$

Then,

$$\text{restab}(\omega^*) \leq \text{restab}(\omega) \tag{6.7}$$

where

$$\text{restab}(\omega) = \left\{ \inf_{\gamma \in (0,1]} \sigma_2 \left(\begin{bmatrix} \Re(\Omega(j\omega)) & -\gamma\Im(\Omega(j\omega)) \\ \gamma^{-1}\Im(\Omega(j\omega)) & \Re(\Omega(j\omega)) \end{bmatrix} \right) \right\} \tag{6.8}$$

Therefore, one has only to check $\omega \in [0, \omega^*]$ to obtain the supremum value in (6.3).

The obtained stability radius from (6.3) provides a basis for assigning eigenvalues for robust stability of systems of DDEs with uncertain parameters.

6.2.2　*Design of robust feedback controller*

In this subsection, an algorithm is presented for the calculation of feedback gains to maintain stability for uncertain systems of DDEs. The approach to eigenvalue assignment using the Lambert W function is used to design robust linear feedback control laws, combined with the stability radius concept. The feedback controller can be designed to stabilize the nominal delayed system (2.1) using the method presented in Chapter 5. However, if the system has uncertainties in the coefficients, which can be introduced by static perturbations of the parameters or can arise in estimating the parameters, the designed controller cannot guarantee stability. Therefore, a *robust* feedback controller is required when uncertainty exists in the parameters. Such a controller can be realized by providing sufficient margins in assigning the rightmost eigenvalues of the delayed system. However, conservative margins over those required can raise problems, such as cost of control. The stability radius, outlined in the previous subsection provides a reasonable measurement of how large the margin should be.

The basic idea of the proposed algorithm is to shift the rightmost eigenvalue to the left by computing the gains in the linear feedback controller in Chapter 5 and increase the stability radius until it becomes larger than the uncertainty of the coefficients. Then, one can obtain a robust controller to guarantee stability of the system with uncertainty.

Algorithm 6.1. Designing a robust feedback controller for systems of DDEs with uncertainty.

Step 1.　Compute the radius, r_1, from actual uncertainties in parameters of given delayed system, (i.e., $r_1 = \sigma_1(\Delta)$).
Step 2.　Using the eigenvalue assignment method presented in Chapter 5, compute $\mathbf{K}$ and $\mathbf{K_d}$, to stabilize the system.
Step 3.　Then, compute the theoretical stability radius of the stabilized system, r_2 from Eq. (6.3).
Step 4.　If $r_1 > r_2$, then, the system can be destabilized by the uncertainties. Therefore, go to Step 2 and increase the margin (compute $\mathbf{K}$ and $\mathbf{K_d}$ to move the rightmost eigenvalues farther to the left).

Example 6.1. From Chapter 5, consider a system

$$\dot{\mathbf{x}}(t) = \begin{bmatrix} 0 & 0 \\ 0 & 1 \end{bmatrix} \mathbf{x}(t) + \begin{bmatrix} -1 & -1 \\ 0 & -0.9 \end{bmatrix} \mathbf{x}(t - 0.1) + \begin{bmatrix} 0 \\ 1 \end{bmatrix} \mathbf{u}(t) \qquad (6.9)$$

Without feedback control, the system in (6.9) has one unstable eigenvalue 0.1098. Using feedback control as in Eq. (5.27), designed by the method presented in Chapter 5, if the desired rightmost eigenvalue is -1, the computed gains are $\mathbf{K} = [-0.1391 - 1.8982]$ and $\mathbf{K_d} = [-0.1236 - 1.8128]$, and the stability radius computed from Eq. (6.2) is 0.6255. However, if the system (6.9) has uncertainties in the parameters

$$\begin{aligned}
\dot{\mathbf{x}}(t) = & \left\{ \begin{bmatrix} 0 & 0 \\ 0 & 1 \end{bmatrix} + \delta\mathbf{A} \right\} \mathbf{x}(t) \\
& + \left\{ \begin{bmatrix} -1 & -1 \\ 0 & -0.9 \end{bmatrix} + \delta\mathbf{A_d} \right\} \mathbf{x}(t - 0.1) + \begin{bmatrix} 0 \\ 1 \end{bmatrix} \mathbf{u}(t)
\end{aligned} \qquad (6.10)$$

and $\sigma([\delta\mathbf{A} \quad \delta\mathbf{A_d}]) = 0.7$, the system can become unstable due to uncertainty. To ensure stability, set the desired rightmost eigenvalue to be -2, then the computed gains are $\mathbf{K} = [0.8805 - 2.1095]$ and $\mathbf{K_d} = [0.9136 - 2.3932]$, and the stability radius in (6.2) increases to 0.9151. Therefore, the system can remain stable despite the uncertainty $(\sigma([\delta\mathbf{A} \quad \delta\mathbf{A_d}]) = 0.7)$. Table 6.1 shows the gains, $\mathbf{K}$ and $\mathbf{K_d}$, corresponding to the several subsets of eigenvalues of $\mathbf{S_0}$.

Table 6.1   The gains, $\mathbf{K}$ and $\mathbf{K_d}$, of the linear feedback controller in (5.27) corresponding to each rightmost eigenvalues. Computed by using the approach for eigenvalue assignment presented in Chapter 5.

Rightmost eigenvalues	K	$\mathbf{K_d}$
-0.5 & -6	[-0.6971 -1.6893]	[-0.7098 -1.5381]
-1.0 & -6	[-0.1391 -1.8982]	[-0.1236 -1.8128]
-1.5 & -6	[-0.3799 -1.6949]	[1.0838 -2.3932]
-2.0 & -6	[0.8805 -2.1095]	[0.9136 -2.3932]
-2.5 & -6	[1.8716 -2.1103]	[0.8229 -2.5904]
-3.0 & -6	[2.5777 -1.7440]	[0.7022 -2.9078]
-3.5 & -6	[2.8765 -1.6818]	[0.9721 -3.1311]
-4.0 & -6	[3.1144 -1.5816]	[1.1724 -3.3304]

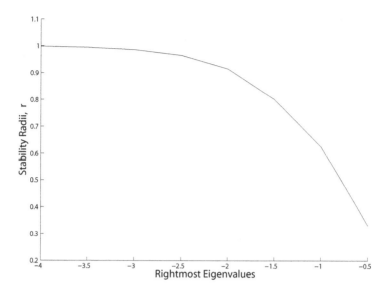

Fig. 6.1 As the eigenvalue moves left, then the stability radius increases consistently, which means, improved robustness.

The computed stability radii versus the rightmost eigenvalues, moving from -0.5 to -4 are shown in Fig. 6.1. As seen in the figure, for the system (6.10), as the eigenvalue moves left, the stability radius increases monotonically. Note that, in general, an explicit relationship between the stability radius and the rightmost eigenvalues is not available, and moving the rightmost eigenvalues further to the left does not always lead to an increase of stability radius (Michiels *et al.*, 2002). However, as shown above, by comparing the stability radius and uncertainty for a given system, Algorithm 6.1 can be used to achieve robust stability of time-delay systems with uncertainty.

Michiels and Roose (2003) developed an algorithm to maximize the stability radius by calculating its sensitivity with respect to the feedback gain for a type of time-delay systems. However, in maximizing it, the rightmost eigenvalues can be moved to undesired positions and one can fail to meet other specifications of the system response. If the system has relatively small uncertainty, instead of maximizing the stability radius, one can focus more on the position of eigenvalues to improve the transient response of the system, which will be discussed in the subsequent section. Also, robust stabilization of systems of DDEs has been investigated by conversion into rational discrete models (Wang *et al.*, 2007), or by canceling undesired dynamics of plants based on system model (de la Sen, 2005). Compared to such methods, the Lambert W function-based approach can improve accuracy and robustness of the controllers.

6.3   Time-Domain Specifications

To meet design specifications in the time-domain, PID-based controllers have been combined with a graphical approach (Shafiei and Shenton, 1997), LQG method using ARE (Suh and Yang, 2005), or Smith predictors (Kaya, 2004). These methods are available for systems with control delays. For systems with state delays, some sufficient conditions based on linear matrix inequality approaches have been proposed (see, e.g., (Mao and Chu, 2006) and the references therein). In this section, the Lambert W function-based approach, presented in Chapter 5, is applied to achieve time-domain specifications via eigenvalue assignment. Unlike other existing methods (e.g., continuous pole placement in (Michiels *et al.*, 2002)), for the first time the Lambert W function-based approach can be used to assign the imaginary parts of system eigenvalues, as well as their real parts, for a critical subset of the infinite eigenspectrum. It is not practically feasible to assign the entire eigenspectrum; however, just by assigning some finite, but rightmost, eigenvalues the transient response of systems of DDEs can be improved to meet time-domain specifications for desired performance.

Example 6.2. Consider the system in (6.9). Table 6.2 shows the gains, $\mathbf{K}$ and $\mathbf{K_d}$, corresponding to the several subsets of eigenvalues of $\mathbf{S_0}$, which have a real part, -0.2, and different imaginary parts, $\pm 0.2i, \pm 0.5i$, and $\pm 1.0i$.

The eigenvalue is written as $\lambda = \sigma \pm \omega_d i = -\zeta \omega_n \pm \omega_n \sqrt{1 - \zeta^2} i$, the requirements for a step response are expressed in terms of the quantities, such as the rise time, t_r, the 1% settling time, t_s, the overshoot, M_p, and the peak time, t_p. In the case of ODEs, if the system is 2nd order without zeros, the quantities have exact representations:

$$t_r = \frac{1.8}{\omega_n}, t_s = \frac{4.6}{\sigma}, M_p = e^{-\pi\zeta/\sqrt{1-\zeta^2}}, t_p = \frac{\pi}{\omega_d} \tag{6.11}$$

For all other systems, however, these provide only approximations, and can only provide a starting point for the design iteration based on the concept

Table 6.2 Gains, **K** and $\mathbf{K_d}$, and parameters corresponding to the several subsets of eigenvalues of $\mathbf{S_0}$.

Rightmost eigenvalues	$-0.2 \pm 0.2i$	$-0.2 \pm 0.5i$	$-0.2 \pm 1.0i$	$-0.5 \pm 1.0i$
σ	-0.2	-0.2	-0.2	-0.5
ω_d	0.2	0.5	1.0	1.0
ω_n	0.2828	0.5385	1.0198	1.1180
$\zeta = \sigma/\omega_n$	0.7	0.3714	0.1961	0.4472
K	[0.0584 -1.7867]	[0.1405 -1.7998]	[0.4311 -1.8152]	[0.2380 -2.1656]
$\mathbf{K_d}$	[0.6789 2.3413]	[0.7802 2.3204]	[1.1421 2.2124]	[0.9027 1.9451]

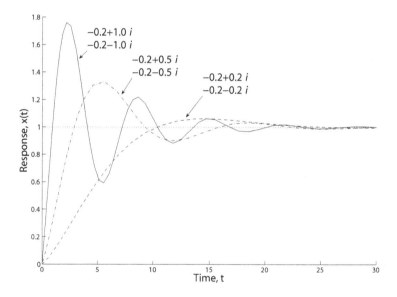

Fig. 6.2 Responses of the system in (6.9) with the feedback (5.27) corresponding to the rightmost eigenvalues in Table 6.2 with different imaginary parts of the rightmost eigenvalues.

of *dominant poles* (Franklin *et al.*, 2005). Figure 6.2 shows the responses corresponding to the rightmost eigenvalues considered in Table 6.2. Not surprisingly, the approximate values from Eq. (6.11) in Table 6.3 are not exactly same as the results obtained from the responses in Fig. 6.2. But,

Table 6.3   Comparison of the actual results for and the approximations using Eq. (6.11) of time-domain specifications for Fig. 6.2.

		Rightmost	$-0.2 \pm 0.2i$	$-0.2 \pm 0.5i$	$-0.2 \pm 1.0i$
t_r	Approximate (Eq. (6.11))		6.4	3.3	1.8
	Actual		6.9	2.5	0.8
t_s	Approximate		23.0	23.0	23.0
	Actual		23.0	23.0	27.0
M_p	Approximate		4.6	28.5	53.4
	Actual		6.0	31.0	75.0
t_p	Approximate		15.7	6.3	3.1
	Actual		14.6	6.0	2.4

for this example, the guidelines for 2nd order ODEs still work well in the case of DDEs. That is, as raising the value of the imaginary part, ω_d, of the rightmost eigenvalue, the rise time, t_r, of system, i.e., the speed at which the system respond to the reference input, decreases from 6.9 to 0.8. On the other hand, the maximum overshoot, M_p, rises from 6% to 75%, which is typically not desirable. In this way, moving up or down the imaginary part, one can adjust the quantities related to the time-domain response and, thus, meet time-domain specifications.

Figure 6.3 shows two responses corresponding to the several subsets of eigenvalues of $\mathbf{S}_0$, which have different real parts (-0.2 and -0.5) with the same imaginary part ($\pm 1.0i$). As seen in the figure, the settling time, the rise time, and overshoot decrease with increasing σ, but the peak time remains almost the same. Thus, for this example, the guidelines based on *dominant poles* for ODEs still work well in case of DDEs. The approach presented in this section is straightforward for systems of ODEs. However, it represents the first approach to assign the real and imaginary parts of the eigenvalues simultaneously to meet time-domain specifications for time delay systems, and is very easy to use, since only the eigenvalues for the principal ($k = 0$) branch are used.

In this approach, it is tried to assign the real and imaginary parts of only the rightmost, thus dominant, eigenvalues. Even though the presented approach handles only a subset of eigenvalues among an infinite eigenspectrum, the subset is rightmost in the complex plane and dominates all other

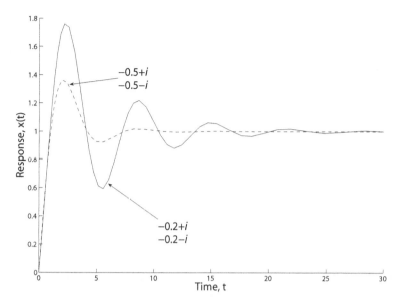

Fig. 6.3 Responses of the system in (6.9) with the feedback (5.27) corresponding to the rightmost eigenvalues in Table 6.3 with different real parts of the rightmost eigenvalues.

eigenvalues. Thus, for other linear time-delay systems with a single delay this method can also be applied to achieve approximate time-domain specifications with linear feedback controllers. The approach presented follows the simple *dominant poles* design guidelines for ODEs, and provides an effective rule of thumb to improve the transient response of systems of DDEs.

6.4 Concluding Remarks

In this chapter, the eigenvalue assignment method based on the Lambert W function is applied to design linear robust feedback controllers and to meet time-domain specifications for LTI systems of DDEs with a single delay. An algorithm for design of feedback controllers to maintain stability for uncertain systems of DDEs is presented. With the algorithm, considering the size of the uncertainty in the coefficients of systems of DDEs via the stability radius, one can find appropriate gains of the linear feedback controller by assigning the rightmost eigenvalues. The procedure presented in this chapter can be applied to uncertain systems, where uncertainty in the system parameters cannot be ignored.

To improve the transient response of time delay systems, the design guideline for systems of ODEs has been used via the Lambert W function-based eigenvalue assignment. The presented approach based on *dominant poles* is quite standard in case of ODEs. However, it has not been previously feasible to use such methods for systems of DDEs. Because, unlike ODEs, DDEs have an infinite number of eigenvalues, and controlling them has not been feasible due to the lack of an analytical solution form.

Using the approach based upon the solution form in terms of the matrix Lambert W function, the analysis for robustness and transient response can be extended from ODEs to DDEs as presented in this chapter. The proposed approach, which is directly related to the position of the rightmost eigenvalues, provides an accurate and effective approach to analyze stability robustness and transient response of DDEs. Even though it is not feasible to assign all of the infinite eigenvalues of time-delay systems, just by assigning the rightmost eigenvalues, which tend to be dominant, one can control systems of DDEs in a way similar to systems of ODEs. This is the advantage of the Lambert W function-based approach over other existing methods.

Chapter 7

Design of Observer-Based Feedback Control for Time-Delay Systems with Application to Automotive Powertrain Control

A new approach for observer-based feedback control of time-delay systems is developed. Time-delays in systems lead to characteristic equations of infinite dimension, making the systems difficult to control with classical control methods. In this chapter, the approach based on the Lambert W function is used to address this difficulty by designing an observer-based state feedback controller via assignment of eigenvalues. The designed observer provides estimation of the state, which converges asymptotically to the actual state, and is then used for state feedback control. The feedback controller and the observer take simple linear forms and, thus, are easy to implement when compared to nonlinear methods. This new approach is applied, for illustration, to the control of a diesel engine to achieve improvement in fuel efficiency and reduction in emissions. The simulation results show excellent closed-loop performance.

7.1 Introduction

As is well known, excellent closed-loop performance can be achieved using state feedback control. In cases where all state variables are not directly measurable, the controller may have to be combined with a state observer, which estimates the state vector. For linear systems, the design of such a controller with an observer is typically carried out based on eigenvalue assignment (Chen, 1984). Unlike systems of linear ordinary differential equations (ODEs), where the methods for eigenvalue assignment are well-developed, the design procedure for linear systems with time-delays in the state variables is not straightforward. In this chapter, a new method for design of observer-based feedback control of time-delay systems is presented, and illustrated with a diesel engine control application.

Successful design of feedback controllers and observers hinges on the ability to check stability and find stabilizing controller and observer gains. In this chapter, a recently developed method using the matrix Lambert

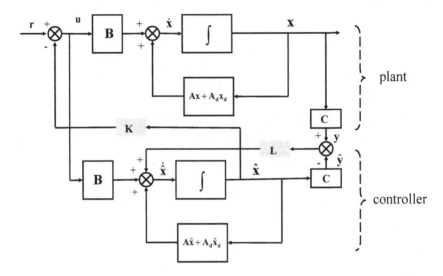

Fig. 7.1   The block diagram for observer-based state feedback control for time-delay systems is analogous to the case for ODEs (Chen, 1984). By choosing gains **K** and **L** an asymptotically stable feedback controller and observer can be designed so that the system has good closed-loop performance using the estimated state variables, $\hat{\mathbf{x}}$, obtained from the system output, **y**.

W function presented in Chapter 5 is applied to design of feedback controllers and observers. The method is used to ensure asymptotic stability and dominant eigenvalues at desired positions in the complex plane to achieve desired performance. Using the Lambert W function-based approach, observer-based controllers for time-delay systems represented by delay differential equations (DDEs) can be designed in a systematic way as for systems of ODEs. That is, for a given time-delay system, the analytical free and forced solutions are derived in terms of parameters of the system (Chapter 2). From the solution form, the eigenvalues are obtained and used to determine stability of the system (Chapter 3). Furthermore, criteria for controllability/observability and Gramians are derived from the solution form (Chapter 4). For a controllable system, a linear feedback controller is designed by assigning dominant eigenvalues to desired locations (Chapter 5), and this can be done to achieve robust stability and to meet time-domain specifications (Chapter 6).

Numerous methods have been developed for control of time-delay systems (e.g., see Table 7.1). However, existing methods enable one to design either the controller or observer, yield nonlinear forms of controllers, and/or do not assign eigenvalues exactly to the desired positions. The approach presented here allows one to design linear feedback controllers and linear observers via eigenvalue assignment (see Fig. 7.1). The observer-based feedback control designed in this chapter offers advantages of accuracy (i.e., no approximation of time-delays), ease of implementation (i.e., no estimates of relevant parameters or use of complex nonlinear controllers), and robustness (i.e., not requiring model-based prediction). For illustration, the developed method is applied to control of a diesel powertrain, where the controller design is challenging due to an inherent time-delay, and the proposed approach can provide advantages in terms of ease of design, as well as the performance of observer-based control. This chapter is organized as follows. In Section 7.2, a problem formulation and background are provided. The proposed method is presented in Section 7.3, and the diesel engine control application is given in Section 7.4. In Section 7.5, a summary and conclusions are presented, and topics for future research are noted.

7.2   Problem Formulation

Consider a real linear time-invariant (LTI) system of DDEs with a single constant time-delay, h, described by Eq. (2.39). With linear state feedback, combined with a reference input, $\mathbf{r}(t) \in R^r$,

$$\mathbf{u}(t) = \mathbf{r}(t) - \mathbf{K}\mathbf{x}(t) - \mathbf{K_d}\mathbf{x}(t - h) \tag{7.1}$$

one can stabilize, improve performance, and/or meet time-domain specifications for the system (2.39) as presented in Chapter 6, under the assumption that all the state variables, $\mathbf{x}(t)$, can be measured directly. This is achieved by choosing $\mathbf{K}$ and $\mathbf{K_d}(\in R^{r \times n})$ based on desired rightmost closed-loop eigenvalues. Note that the Lambert W function-based approach is applicable to systems with a single delay as in Eq. (2.39). For systems with multiple delays caused by, e.g., additional feedback delays or delays in sensors, stability results introduced in (Olgac *et al.*, 2005) can be applied.

In cases where direct access to all state variables is limited, use of a state observer (estimator) is needed to obtain $\hat{\mathbf{x}}(t)$, an estimate of the state variable, $\mathbf{x}(t)$. Like systems of ODEs, an asymptotically stable closed-loop system with a state observer (so-called Luenberger observer where $\hat{\mathbf{x}}(t)$ converges asymptotically to $\mathbf{x}(t)$ as t goes to infinity) can be achieved by placing eigenvalues for the observer dynamics at desired locations in the complex plane (e.g., on the left half plane (LHP)). However, in contrast

Table 7.1   Motivation for a new approach for design of observers: for time-delay systems, various studies devoted to observer design are summarized in this table.   For comparisons of approaches for feedback controllers, refer to (Richard, 2003; Yi et al., 2010b).

Description of approach		References
Spectral decomposition-based methods	Observer: an integro-differential form	[Bhat and Koivo (1976b)] [Pearson and Fiagbedzi (1989)] [Salamon (1980)]
	Observer: a linear form	[Leyvaramos and Pearson (1995)]
Lyapunov framework	Linear matrix inequality (LMI)	[Darouach (2001)] [Bengea et al. (2004)]
	Algebraic Riccati equation (ARE)	[Pila et al. (1999)]
Coordinate transformation		[Trinh (1999)]
Finite spectrum assignment (FSA)		[Jankovic and Kolmanovsky (2009)]

to ODEs, systems of DDEs, as in Eq. (2.39), have an infinite number of eigenvalues (e.g., see Fig. 2.1) and, thus, calculation and assignment of all of them is not feasible.

The state estimation problem for time-delay systems has been a topic of research interest (e.g., see (Bengea et al., 2004) and the references therein for a survey). The problem has been approached by using methods based on spectral decomposition and state transformation developed in (Bhat and Koivo, 1976b), (Pearson and Fiagbedzi, 1989), and (Salamon, 1980). Such methods require extensive numerical computations to locate the eigenvalues of time-delay systems. Prediction-based approaches (e.g., FSA) with a coordinate transformation have been used to address this type of problem in (Jankovic and Kolmanovsky, 2009). Converting time-delay systems into non-delay ones, the observer of an integro-differential form is designed to assign the eigenvalues of finite dimensional systems. Based on the assumption that stabilizing feedback gains exist and are known, an observer can be designed based on a coordinate transformation (Trinh, 1999). For such an approach, it is assumed that the system is stabilized by the memoryless linear state feedback, $\mathbf{u}(t) = -\mathbf{K}\mathbf{x}(t)$, and the gain, $\mathbf{K}$, is known. Also, Lyapunov-based approaches have been used for development of design methods for observers and/or controllers (e.g., ARE (Pila et al., 1999), LMI (Bengea et al., 2004; Darouach, 2001)). See the comparison of various developed approaches in Table 7.1.

7.2.1 *Eigenvalue assignment*

While problems in handling time-delay systems arise from the difficulty in: 1) checking the stability and 2) finding stabilizing gains, Eq. (5.16) for eigenvalue assignment provides a explicit formulation useful to address such problems, as shown in (Yi *et al.*, 2010b) and (Yi *et al.*, 2008c) with numerical examples. However, as explained in Subsection 7.2.2 below, due to assignability issues, Eq. (5.16), which is solved by using numerical nonlinear equation solvers (e.g., *fsolve* in Matlab), may not always yield a solution for $\mathbf{K}$ and $\mathbf{K_d}$. To resolve such a problem, instead of using Eq. (5.16), one can try with fewer desired eigenvalues, or with just the real parts of the desired eigenvalues (Yi *et al.*, 2010b) to reduce the number of constraints, as

$$\lambda_{rm}(\mathbf{S}_0) = \lambda_{des} \qquad (a)$$
$$\Re\{\lambda_{rm}(\mathbf{S}_0)\} = \Re\{\lambda_{des}\} \quad (b)$$

$$(7.2)$$

where $\lambda_{rm}(\mathbf{S}_0)$ are the rightmost eigenvalues from among the n eigenvalues $\mathbf{S}_0$, and $\Re$ indicates the real part of its argument. In numerical computation one can use, for example, functions in Matlab, such as *max* and *real*.

In Chapter 5, the method for eigenvalue assignment, based on the *Conjecture* in Subsection 3.3.1, was used to design only full-state feedback control as in Eq. (7.1) (Yi *et al.*, 2010b). In this chapter, it is now used to find the controller and observer gains (i.e., $\mathbf{K}$ and $\mathbf{L}$ in Eqs. (7.3) and (7.4) in the next section) for the combined observer-based control in Fig. 7.1. This is described in Section 7.3 below. In other words, using Eq. (5.16), or Eq. (7.2), it will be shown that one can assign both controller and observer rightmost (i.e., dominant) poles for the infinite dimensional closed-loop eigenspectrum of the observer-based controller for time-delay systems shown in Fig. 7.1.

7.2.2 *Controllability, observability, and eigenvalue assignability*

For systems of DDEs, controllability and observability have been studied extensively (see e.g., (Richard, 2003), (Yi *et al.*, 2008a) and the references therein). Unlike systems of ODEs, there exist numerous different definitions of controllability and observability for systems of DDEs depending on the nature of the problem under consideration (e.g., approximate, spectral, weak, strong, point-wise and absolute controllability). Among the various notions, point-wise controllability and point-wise observability were investigated to derive criteria and Gramians for such properties using the solution form in Eq. (2.43) based on the Lambert W function in Chapter 4.

For linear systems of ODEs, controllability (or observability) is equivalent to the arbitrary assignability of the eigenvalues of the controller (observer) (Chen, 1984). However, conditions for such arbitrary assignment are still lacking for systems of DDEs. Even for the scalar case of Eq. (2.1), limits in arbitrary assignment of eigenvalues exist and depend on the values of the time-delay and the coefficients (see Appendix C). Although, for a simple scalar DDE, study of the limits has been conducted in (Beddington and May, 1975; Bellman and Cooke, 1963; Cooke and Grossman, 1982), generalization of such results are challenging. The relationship between the derived criteria for controllability and eigenvalue assignability by using a 'linear feedback controller' was partially studied in (Yi *et al.*, 2010b) with examples, and is being further investigated by the authors. Although extensive research during recent decades has been reported in the literature, the relationship between eigenvalue assignment and controllabilty/observabilty is still an open research problem.

7.3   Design of Observer-Based Feedback Controller

This section describes a systematic design approach for the combined controller-observer for time-delay systems (see Fig. 7.1). The observer estimates the system states from the output variables, while the control provides inputs to the system as a linear function of the estimated system states.

Step 1.   Obtain the equation for the closed-loop system with $\mathbf{K}$ and $\mathbf{K_d}$:

$$\dot{\mathbf{x}} = \mathbf{Ax} + \mathbf{A_d x_d} + \mathbf{Bu}$$
$$\mathbf{u} = -\mathbf{Kx} - \mathbf{K_d x_d} + \mathbf{r} \tag{7.3}$$
$$\Rightarrow \dot{\mathbf{x}} = (\mathbf{A} - \mathbf{BK})\,\mathbf{x} + (\mathbf{A_d} - \mathbf{BK_d})\,\mathbf{x_d} + \mathbf{Br}$$

where $\mathbf{x_d} \equiv \mathbf{x}(t-h)$. Then, choose the desired positions of the *rightmost* eigenvalues of the feedback controller dynamics. They can be selected, for example, to meet design specifications in the time domain with desired damping ratio, ζ, desired natural frequency, ω_n, of the closed loop response and as $\lambda_{des} = \sigma \pm j\omega_d = -\zeta\omega_n \pm j\omega_n\sqrt{1 - \zeta^2}$.

Step 2. Using the desired eigenvalues, Eq. (5.16) and Eq. (2.20), the gains, $\mathbf{K}$ and $\mathbf{K_d}$, are obtained numerically for a variety of initial conditions by an iterative trial and error procedure with the coefficients of the closed-loop system in Eq. (7.3), $\mathbf{A}' \equiv \mathbf{A} - \mathbf{BK}$ and $\mathbf{A_d'} \equiv \mathbf{A_d} - \mathbf{BK_d}$. As explained in Subsection 7.2.1, if solutions cannot be found, one can try with fewer desired eigenvalues, or with just the real parts of the desired rightmost eigenvalues (i.e., using Eq. (7.2) instead of Eq. (5.16)), to find the control gains.

Step 3. Consider an observer with gain $\mathbf{L}$:

$$\dot{\mathbf{x}} = \mathbf{Ax} + \mathbf{A_d x_d} + \mathbf{Bu}$$
$$\mathbf{y} = \mathbf{C}\mathbf{x(t)}$$
$$\dot{\hat{\mathbf{x}}} = \mathbf{A\hat{x}} + \mathbf{A_d \hat{x}_d} + \mathbf{L(y - C\hat{x})} + \mathbf{Bu} \qquad (7.4)$$
$$\dot{\mathbf{x}} - \dot{\hat{\mathbf{x}}} = \mathbf{A(x - \hat{x})} + \mathbf{A_d (x_d - \hat{x}_d)} - \mathbf{L(y - C\hat{x})}$$
$$\Rightarrow \dot{\mathbf{e}} = (\mathbf{A - LC})\mathbf{e} + \mathbf{A_d e_d}$$

where $\mathbf{e} \equiv \mathbf{x} - \hat{\mathbf{x}}$. Then, choose the desired positions of the *rightmost* eigenvalues of the observer dynamics. A reasonable choice of desired positions of observer rightmost eigenvalues mainly depends on the amount of measurement noise and the size of modeling inaccuracies. While fast eigenvalues amplify measurement noise, slow eigenvalues lead to slow convergence of the estimates of the state variables. A typical "rule of thumb" is that the magnitudes of the negative real parts of the rightmost eigenvalues of (7.4) should be $1.5 \sim 2$ times larger than those of Eq. (7.3) to guarantee fast response (Chen, 1984).

Step 4. Using the desired eigenvalues, Eq. (5.16) and Eq. (2.20), find the observer gain, $\mathbf{L}$, with new coefficients in Eq. (7.4), $\mathbf{A}' \equiv \mathbf{A} - \mathbf{LC}$ and $\mathbf{A_d'} \equiv \mathbf{A_d}$. As in Step 2, if solutions cannot be found, one can try with fewer desired eigenvalues, or with just the real parts of the desired rightmost eigenvalues (i.e., using Eq. (7.2) instead of Eq. (5.16)).

In Step 1, the desired positions of the rightmost eigenvalues, with selected damping ratio and frequency, provide only an approximation for systems of order higher than two. They provide starting points for the design iteration based on the concept of *dominant poles* (Yi *et al.*, 2008c; Franklin *et al.*, 2005).

Note that, as mentioned in Subsection 7.2.2, unlike ODEs, conditions for assignability are still lacking for systems of DDEs. Even for the scalar case of Eq. (2.1), limits in arbitrary assignment of eigenvalues exist (e.g., see Appendix C of this chapter). For systems of DDEs, depending on the structure or parameters of the given system, there exists limitations on the rightmost eigenvalues. In that case, the above approach does not yield any solution for the controller and observer gains. To resolve the problem, one can find the gains by using a trial and error method with fewer desired eigenvalues (or with just the real parts of the desired eigenvalues), or different values of the desired rightmost eigenvalues as explained in Steps 2 and 4.

Although the Kalman filter for time-delay systems renders an observer optimal in the sense of minimizing the estimation error in the presence of noise and model uncertainty (see, e.g., (Fattouh *et al.*, 1998; Yu, 2008) and the references therein), such an approach requires the selection of covariance matrices for process and measurement noise by trial-and-error to obtain the desired performance of the filter/estimator. On the other hand, the design of observers via eigenvalue assignment may be sub-optimal, but can achieve a similar performance by adjusting desired location for eigenvalues (Chen, 1984).

Even though only an approach for the full-dimensional observer is presented here, a reduced-dimensional observer, if needed, can be designed in a similar way.

7.3.1   *Separation principle*

For systems of ODEs, it is shown that the eigenvalues of the state estimator are not affected by the feedback and, consequently, the design of state feedback and the design of the state estimator can be carried out independently (i.e., the so-called separation principle). For the time-delay system in (7.3) and (7.4), it can be shown in a straightforward manner that the separation principle holds. The two equations can be combined into

$$
\begin{bmatrix} \dot{\mathbf{x}} \\ \dot{\mathbf{e}} \end{bmatrix} = \begin{bmatrix} (\mathbf{A} - \mathbf{BK} - \mathbf{BK_d}) & \mathbf{BK} \\ \mathbf{0} & (\mathbf{A} - \mathbf{LC}) \end{bmatrix} \begin{bmatrix} \mathbf{x} \\ \mathbf{e} \end{bmatrix} + \begin{bmatrix} \mathbf{A_d} & \mathbf{0} \\ \mathbf{0} & \mathbf{A_d} \end{bmatrix} \begin{bmatrix} \mathbf{x_d} \\ \mathbf{e_d} \end{bmatrix} \quad (7.5)
$$

Then, the eigenvalues are roots of the characteristic equation given by

$$
\det \begin{bmatrix} s\mathbf{I} - \mathbf{A} + \mathbf{BK} + \mathbf{BK_d} - \mathbf{A_d}e^{-sh} & -\mathbf{BK} \\ \mathbf{0} & s\mathbf{I} - \mathbf{A} + \mathbf{LC} - \mathbf{A_d}e^{-sh} \end{bmatrix}
$$
$$
= 0
$$
$$
\Rightarrow \det \begin{bmatrix} s\mathbf{I} - \mathbf{A} + \mathbf{BK} + \mathbf{BK_d} - \mathbf{A_d}e^{-sh} \end{bmatrix} \times
$$
$$
\det \begin{bmatrix} s\mathbf{I} - \mathbf{A} + \mathbf{LC} - \mathbf{A_d}e^{-sh} \end{bmatrix} = 0
$$
$$
(7.6)
$$

Therefore, the two sets of eigenvalues can be specified separately and the introduction of the observer does not affect the eigenvalues of the controller. Hence, selection of gains $\mathbf{K}$ (and/or $\mathbf{K_d}$) and $\mathbf{L}$ can be performed independently.

7.4 Application to Diesel Engine Control

In this section, control of a diesel engine is considered to illustrate the advantages and potential of the method proposed in Section 7.3. Specifically, a feedback controller and observer are designed via eigenvalue assignment using the Lambert W function-based approach. A diesel engine with an exhaust gas recirculation (EGR) valve and a turbo-compressor with a variable geometry turbine (VGT) was modeled in (Jankovic and Kolmanovsky, 2000) with 3 state variables, $\mathbf{x}(t) \equiv \{x_1 \ x_2 \ x_3\}^T$: intake manifold pressure (x_1), exhaust manifold pressure (x_2), and compressor power (x_3). The model includes intake-to-exhaust transport delay ($h = 60\text{ms}$ when engine speed, N, is 1500 RPM). Thus, a linearized system of DDEs was introduced in (Jankovic and Kolmanovsky, 2009), for a specific operating point ($N = 1500$ RPM) whose coefficients are given by:

$$\mathbf{A} = \begin{bmatrix} -27 & 3.6 & 6 \\ 9.6 & -12.5 & 0 \\ 0 & 9 & -5 \end{bmatrix}, \mathbf{A_d} = \begin{bmatrix} 0 & 0 & 0 \\ 21 & 0 & 0 \\ 0 & 0 & 0 \end{bmatrix}, \mathbf{B} = \begin{bmatrix} 0.26 & 0 \\ -0.9 & -0.8 \\ 0 & 0.18 \end{bmatrix} \quad (7.7)$$

Because of the time-delay, which is caused by the fact that the gas in the intake manifold enters the exhaust manifold after transport time, h, the system can be represented by a system of DDEs as in Eq. (2.1) with the coefficients in Eq. (7.7). The number of eigenvalues is infinite and one of them is positive real. Thus, the response of this linearized system shows unstable behavior (see the eigenspectrum and the response in Fig. 7.2). The system with the coefficients in (7.7) satisfies the conditions for *pointwise controllability* and *observability*. That is, all rows of (4.11) and all columns of (4.14) with the coefficients in (7.7) are linearly independent. Therefore, the system of (7.7) is *pointwise controllable* and *pointwise observable* (Yi et al., 2008a) (see Chapter 4).

For a non-delay model, which is also unstable, by constructing the control Lyapunov function (CLF) a feedback control law was designed in (Jankovic and Kolmanovsky, 2000). A feedback controller for the diesel engine systems with the time-delay was developed in (Jankovic, 2001) using

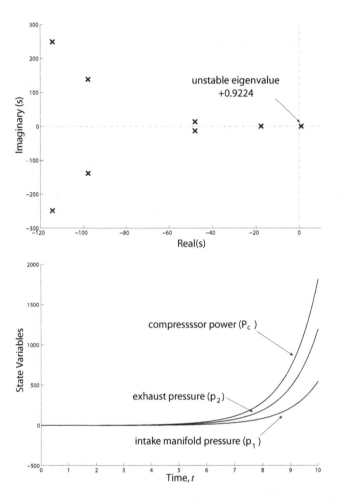

Fig. 7.2 Eigenspectrum of the linearized diesel engine system with the coefficients in (7.7). Due to the inherent transport time-delay the number of eigenvalues is infinite and without any control the system has an unstable eigenvalue (left). Thus, the response of the system is unstable (right) (Jankovic and Kolmanovsky, 2009).

the concept of control Lyapunov-Krasovsky functionals (so-called CLKF). In (Jankovic and Kolmanovsky, 2009), FSA combined with a coordinate transformation (FSA cannot be applied directly to the system where time-delays are not in actuation) was used for design of the observer for the system, which takes an integro-differential form.

As shown in Subsection 7.3.1, the separation principle holds, and, thus the controller and observer in Fig. 7.3 can be designed independently.

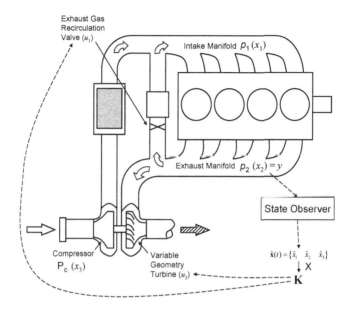

Fig. 7.3 Diagram of a diesel engine system in Eq. (7.7). With limited measurement $y = x_2$, all the state variables are estimated by using an observer, and then used for state feedback control (Jankovic and Kolmanovsky, 2009).

First consider the case of linear full-state feedback. The control input, $\mathbf{u}(t) = \{u_1(t)\ u_2(t)\}^T$, where $u_1(t)$ is a control input for EGR valve opening and $u_2(t)$ is a control input for the turbine (VGT) mass flow rate (for a detailed explanation, the reader may refer to (Jankovic and Kolmanovsky, 2000)), is given by

$$\mathbf{u}(t) = \mathbf{K}\mathbf{x}(t) + \mathbf{r}(t) \tag{7.8}$$

where $\mathbf{K}$ is the 2×3 feedback gain matrix,

$$\mathbf{K} = \begin{bmatrix} k_1 & k_2 & k_3 \\ k_4 & k_5 & k_6 \end{bmatrix} \tag{7.9}$$

Then, the system matrices of the closed-loop system become

$$\mathbf{A}' \equiv \mathbf{A} + \mathbf{B}\mathbf{K} =$$

$$\begin{bmatrix} -27 + 0.26k_1 & 3.6 + 0.26k_2 & 6 + 0.26k_3 \\ 9.6 - 0.9k_1 - 0.8k_4 & -12.5 - 0.9k_2 - 0.8k_5 & 0 - 0.9k_3 - 0.8k_6 \\ 0.18k_4 & 9 + 0.18k_5 & -5 + 0.18k_6 \end{bmatrix}, \tag{7.10}$$

$$\mathbf{A}'_{\mathbf{d}} \equiv \mathbf{A_d}$$

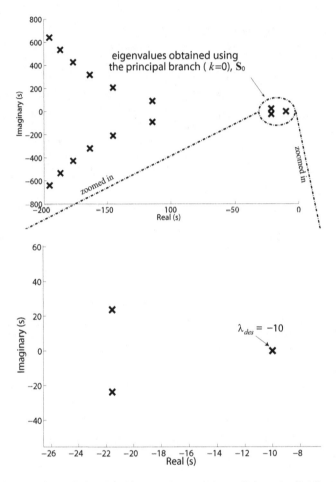

Fig. 7.4   Eigenvalues of the closed-loop systems with coefficients in (7.10) and (7.11). Among an infinite number of eigenvalues, the rightmost (i.e., dominant) subset is computed by using the principal branch ($k = 0$) and all the others are located to the left. Note that the rightmost eigenvalue is placed exactly at the desired position, $\lambda_{des} = -10$, and the open-loop unstable system is stabilized (compare to Fig. 7.2).

Following the Steps 1 and 2 introduced in Section 7.3, based on the Lambert W function, the gain, $\mathbf{K}$, is selected so that the system can have improved performance as well as be stabilized. For the system in Eq. (7.7), it was not possible to assign all eigenvalues of $\mathbf{S}_0$ by using Eq. (5.16). Instead, by reducing the number of eigenvalues specified to one, one can find the feedback gain and assign the rightmost eigenvalue of the system with Eq. (7.2-a).

For example, assuming that the desired rightmost eigenvalue, $\lambda_{des} = -10$, which is chosen by considering the desired speed of the closed-loop system (Jankovic and Kolmanovsky, 2009), the resulting feedback gain is obtained as

$$\mathbf{K} = \begin{bmatrix} 0.0001 & -13.8835 & 0 \\ 0 & 50.3377 & 50.8222 \end{bmatrix} \tag{7.11}$$

The resulting eigenspectrum is shown in Fig. 7.4. Among an infinite number of eigenvalues, the rightmost (i.e., dominant) subset is computed by using the principal branch ($k = 0$) and all the others are located to the left, which is one of the prominent advantages of the Lambert W function-based approach. Note that the rightmost eigenvalue is placed exactly at the desired position, $\lambda_{des} = -10$, and the unstable system is stabilized (compare to the eigenspectrum in Fig. 7.2).

Next consider the design of an observer to estimate the unmeasured states. The observer gain, $\mathbf{L}$, is obtained in a similar way, following Steps 3 and 4 in Section 7.3. Considering available sensors (Jankovic and Kolmanovsky, 2009), only the exhaust manifold pressure is measured. Thus, $y = x_2$ and the output matrix and observer gain are given by

$$\mathbf{C} = \begin{bmatrix} 0 & 1 & 0 \end{bmatrix}, \mathbf{L} = \begin{bmatrix} L_1 \\ L_2 \\ L_3 \end{bmatrix} \tag{7.12}$$

Then, the new coefficients of the dynamical equation of the state observer become

$$\mathbf{A} - \mathbf{LC} = \begin{bmatrix} -27 & 3.6 - L_1 & 6 \\ 9.6 & -12.5 - L_2 & 0 \\ 0 & 9 - L_3 & -5 \end{bmatrix}, \mathbf{A_d} = \begin{bmatrix} 0 & 0 & 0 \\ 21 & 0 & 0 \\ 0 & 0 & 0 \end{bmatrix} \tag{7.13}$$

For example, for a chosen desired eigenvalue $\lambda_{des} = -15$ so that the dynamics of the observer is well damped and faster than the controller dynamics, the gain obtained is

$$\mathbf{L} = \begin{bmatrix} 6.4729 \\ 9.5671 \\ 16.0959 \end{bmatrix} \tag{7.14}$$

Similarly to the controller case, one can find the observer gain, $\mathbf{L}$, just with one rightmost eigenvalue specified (i.e., by using Eq. (7.2-a)). The resulting eigenspectrum is shown in Fig. 7.5.

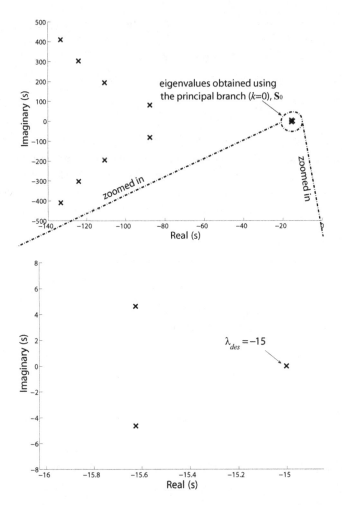

Fig. 7.5   Eigenvalues of the closed-loop systems with coefficients in (7.13) and (7.14): among an infinite number of them, the rightmost (i.e., dominant) subset is computed by using the principal branch ($k = 0$) and all the others are located to the left. Note that the rightmost eigenvalue is placed exactly at the desired position, $\lambda_{des} = -15$, and the unstable system is stabilized (compare to Fig. 7.2).

As mentioned in Sections 7.2 and 7.3 and Appendix C, there exist limits in assignment of eigenvalues with linear controllers or observers. For the system with coefficients in Eq. (7.7), the rightmost eigenvalues can be moved as far to the left as $\lambda_{des} = -25.0$ for the controller and $\lambda_{des} = -15.3$ for the observer, respectively, and the corresponding limiting gains,

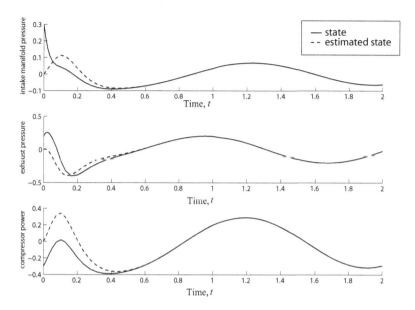

Fig. 7.6 Responses with each controller: the reference inputs for this simulation run are selected randomly: r_1 is a step input with amplitude 0.5 and r_2 is a sine wave with amplitude 20 and frequency 0.7 (Hz). The rightmost eigenvalues of feedback control and observer are -10 and -15, respectively. The state variables estimated by using observer (dashed line) converge into those of the plant (solid line), which are stabilized by state feedback control.

as determined numerically for this example, are:

$$\mathbf{K} = \begin{bmatrix} -0.0004 & -3.3044 & -0.0006 \\ 0.0020 & 45.9131 & 84.0699 \end{bmatrix}, \quad \mathbf{L} = \begin{bmatrix} 6.4650 \\ 9.5660 \\ 16.0991 \end{bmatrix} \tag{7.15}$$

Overall performance of the controlled system in Fig. 7.1 with the parameters in (7.7), (7.11), (7.12) and (7.14) is shown in Fig. 7.6. The reference inputs for this simulation run are selected arbitraily: r_1 is a step input with amplitude 0.5 and r_2 is a sine wave with amplitude 20 and frequency 0.7 (Hz). The rightmost eigenvalues of the feedback controller and observer are -10 and -15, respectively. The state variables estimated using the observer (dashed line) converge to those of the plant (solid line), which are stabilized by state feedback control.

As shown in Fig. 7.1, the asymptotically stable feedback controller with observer takes a simple form similar to that for linear systems of ODEs (Chen, 1984). They do not require, as with previous approaches, the approximate integration of state variables during finite intervals, nor construction of a cost function or inequalities. This can lead to ease of design, analysis, and implementation, which is one of the main advantages of the proposed approach.

7.5 Conclusions

In this chapter, a new approach for the design of observer-based state feedback control for time-delay systems was developed. The separation principle is shown to hold, thus, the controller gain and the observer gain can be independently selected so that the two dynamics are simultaneously asymptotically stable. The design hinges on eigenvalue assignment to desired locations that are stable, or ensure a desired damping ratio and natural frequency. The main difficulty, which is addressed in this chapter, is caused by the fact that systems of DDEs have an infinite number of eigenvalues, unlike systems of ODEs. Thus, to locate them all to desired positions in the complex plane is not feasible. To find the dominant subset of eigenvalues and achieve desired eigenvalue assignment, the Lambert W function-based approach, developed recently by the authors has been used. Using the proposed approach, the feedback and the observer gains are obtained by placing the rightmost, or dominant, eigenvalues at desired values. The designed observer provides an estimate of the state variables, which converges asymptotically to the actual state and is then used for state feedback to improve system performance. The technique developed is applied to a model for control of a diesel engine, and the simulation results show excellent performance of the designed controller and observer.

The proposed method complements existing methods for observer-based controller design and offers several advantages. The designed observer-based controller for DDEs has a linear form analogous to the usual case for ODEs. The rightmost (i.e., dominant) eigenvalues, for both observer and controller, are assigned exactly to desired feasible positions in the complex plane. The designed control can have improved accuracy by not ignoring or approximating time-delays, ease of implementation compared to nonlinear forms of controllers, and robustness since it does not use model-based prediction.

To make the proposed approach more effective, the relation between controllability (or observability) and eigenvalue assignability for time-delay systems needs to be investigated further. Specifically, research is needed to generalize the results for the scalar case (see Appendix C). As new engine technologies are continuously developed, the proposed design approach can play a role in handling delay problems for automotive powertrain systems. Other than the presented diesel control, for example, air-to-fuel ratio control, where time-delays exist due to the time between fuel injection and sensor measurement for exhaust, and idle speed control, where time-delays exist due to the time between the intake stroke of the engine and torque production, are also being studied.

Chapter 8

Eigenvalues and Sensitivity Analysis for a Model of HIV Pathogenesis with an Intracellular Delay

During the past decade, numerous studies have aimed at better understanding of the human immunodeficiency virus (HIV). For example, the combination of mathematical modeling and experimental results has made a significant contribution. However, time-delays, which play a critical role in various biological models including HIV models, are still not amenable to many traditional analysis methods. In this chapter, a recently developed approach using the Lambert W function is applied to handle the time-delay inherent in an HIV pathogenesis dynamic model. Dominant eigenvalues in the infinite eigenspectrum of these time-delay systems are obtained and used to understand the effects of the parameters of the model on the immune system. Also, the result is extended to analyze the sensitivity of the eigenvalues with respect to uncertainty in the parameters of the model. The research makes it possible to know which parameters are more influential relative to others, and the obtained information is used to make predictions about HIV's outcome.

8.1 Introduction

During the past decade, a number of mathematical models for the human immunodeficiency virus (HIV), based on systems of differential equations, have been developed. These models, combined with experimental results, have yielded important insights into HIV pathogenesis (Adams *et al.*, 2005; Kirschner and Webb, 1996; Nowak *et al.*, 1997; Perelson and Nelson, 1999; Perelson, 2002). This success in modeling the HIV pathogenesis dynamics has led to various analyses (Banks and Bortz, 2005; Bortz and Nelson, 2006, 2004; Nelson and Perelson, 2002), and helped in designing better therapy regimes (Adams *et al.*, 2004).

To account for the time between viral entry into a target cell and the production of new virus particles, models that include time-delays have been introduced (Herz *et al.*, 1996; Mittler *et al.*, 1998; Nelson *et al.*, 2000). Models of HIV infection that include intracellular delays are more accurate representations of the biology and change the estimated values of kinetic parameters when compared to models without delays (Nelson and Perelson, 2002). Also, it has been shown that allowing for time-delays in the model better predicts viral load data when compared to models with no time-delays (Bortz and Nelson, 2006; Ciupe *et al.*, 2006; Nelson *et al.*, 2001). Due to the complexity of delay differential equations (DDEs), many scientists do not include them in their models. However, many biological processes have inherent delays and including them may lead to additional insights in the study of complicated biological processes (Nelson and Perelson, 2002).

In this chapter, dominant eigenvalue analysis and its sensitivity with respect to parameters in the model of HIV dynamics are studied. For this research, the matrix Lambert W function approach is applied to investigate analytically the HIV pathogenesis model with an intracellular delay. Eigenvalues of the delayed systems are obtained and used: i) to analyze the effects of time-delay on the stability and decay rate of the viral load, and ii) to determine the stability of the immune systems. Also, via sensitivity analysis of the eigenvalues with respect to parameters, the effects of parameters are studied. The approach presented in this chapter for HIV dynamics can be used to deal with time-delay terms in many other pathogenesis models (e.g., hepatitis B viral dynamics (Ciupe *et al.*, 2007) and tuberculosis (Marino *et al.*, 2007)).

8.2   HIV Pathogenesis Dynamic Model with an Intracellular Delay

The HIV pathogenesis dynamic models have been used to interpret experimental results for complex immune systems. Research on relations between parameters in the models and their impact on the immune system has been

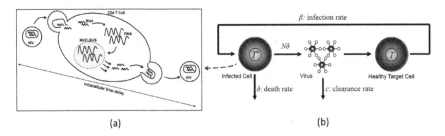

(a)   (b)

Fig. 8.1   HIV-1 infects target cells (T) with a rate constant β and causes them to become productively infected cells (T^*) (b). Time-delay results from the time between initial viral entry into a cell and subsequent viral production (a).

reported in the literature (see, e.g., (Perelson and Nelson, 1999; Perelson, 2002)) and have made a significant contribution during the past decade. When an intracellular delay is included, the models of HIV infection provide more accurate representations of the biology. This is because allowing for time-delays in the model enables it to better predict the viral load dynamics. One of the delay models, where it is assumed that the generation of virus producing cells at time t is due to the infection of target cells at time $t - h$ as seen in Fig. 8.1, consists of systems of coupled delay differential equations given by (Nelson *et al.*, 2000):

$$\frac{dT^*(t)}{dt} = \beta T_0 V_I(t - h) - \delta T^*(t)$$

$$\frac{dV_I(t)}{dt} = (1 - n_p)N\delta T^*(t) - cV_I(t) \qquad (8.1)$$

$$\frac{dV_{NI}(t)}{dt} = n_p N\delta T^*(t) - cV_{NI}(t)$$

where t is the elapsed time since treatment was initiated (i.e., $t = 0$ is the time of onset of the drug effect), and T^* is the concentration of productively infected T-cells. The state variables V_I and V_{NI} represent the plasma concentrations of virions in the infectious pool (produced before the drug effect) and in the noninfectious pool (produced after the drug effect), respectively. In Eq. (8.1), it is assumed HIV infects target cells with a rate β and causes them to become productively infected T-cells, T^*. The time-delay, h, in Eq. (8.1) results from the time between initial viral entry into a cell and subsequent viral production, and is termed "intracellular delay". In this model, c is the rate for virion clearance; δ is the rate of loss of the virus-producing cell; N is the number of new virions produced per infected cell during its lifetime; T_0 is the target T-cell concentration; n_p represents the drug efficacy of a protease inhibitor, a drug that inhibits the cleaving of viral polyproteins and renders newly produced virions non-infectious, V_{NI}. The term $(1 - n_p)$ represents the level of leakiness of a protease inhibitor and if $n_p = 1$, the protease inhibitor is 100% effective and no infectious virus particles are produced. The parameters in (8.1) have been estimated by applying the models to data from drug perturbation experiments (Nelson *et al.*, 2001). For the research presented in this chapter, the parameter set for patient 103, which is given in Table 8.1, is used (Nelson *et al.*, 2001). Viral load, $V_I + V_{NI}$, had been collected from patient 103 after administration (600 mg twice daily) of a potent inhibitor (Ritonavir) of HIV protease. For detailed study, refer to (Perelson *et al.*, 1996) on the experiment and the data.

Table 8.1   Estimated parameter values from one (patient 103) of the 5 patients studied in (Nelson *et al.*, 2001).

Name	Description	Value	Reference
T_0	Target T-cell concentration	$408\ cells\ mm^{-3}$	(Ho *et al.*, 1995)
h	Intracellular delay	$0.91\ days$	(Nelson *et al.*, 2000)
δ	Death rate of an infected T-cell	$1.57/day$	(Perelson *et al.*, 1996)
c	Clearance rate of virus	$4.3/day$	(Perelson *et al.*, 1996)
N	Bursting term for viral production after lysis	$480\ virions/cells$	(Perelson *et al.*, 1996)
n_p	Protease inhibitor efficacy	0.7	(Perelson and Nelson, 1999)
β	Viral infectivity rate	$\dfrac{c}{NT_0}$	(Ho *et al.*, 1995)

The system in Eq. (8.1) is expressed in the form of Eq. (2.1) with the coefficients

$$\mathbf{A} = \begin{bmatrix} -\delta & 0 & 0 \\ (1 - n_p)N\delta & -c & 0 \\ n_p N\delta & 0 & -c \end{bmatrix}, \quad \mathbf{A_d} = \begin{bmatrix} 0 & \beta T_0 & 0 \\ 0 & 0 & 0 \\ 0 & 0 & 0 \end{bmatrix} \tag{8.2}$$

and the initial conditions $\mathbf{g}(t) = \mathbf{x}_0 = \{T_{ss}^* \quad V_{ss} \quad 0\}^T$. The characteristic equation of the system in Eq. (8.1) is derived as

$$H(\lambda) = \left\{ \lambda^2 + (\delta + c)\lambda + \delta c - (1 - n_p)\delta N\beta T_0 e^{-\lambda h} \right\} (\lambda + c) \tag{8.3}$$

And from the roots of Eq. (8.3), the eigenvalues, λ, of the system (8.1) are obtained. Due to the term, $e^{-\lambda h}$, Eq. (8.3) becomes infinite-dimensional and, thus, an infinite number of roots satisfy the equation. The principal difficulty in studying DDEs results from this special transcendental character, and the determination of this spectrum typically requires numerical, approximate, and graphical approaches (Richard, 2003). Computing, analyzing, and controlling the infinite eigenspectrum are not as straightforward as for systems of ordinary differential equations (ODEs). Instead, for the time-delay system in Eq. (8.1), it is crucial to compute and analyze the dominant eigenvalues. To do that, the Lambert W function-based approach (e.g., see Sections 2.2 and 3.3.1) is applied.

8.3   Rightmost Eigenvalue Analysis

For the model of HIV in (8.1), the stability of a patient's immune system and the viral decline rate can be expressed with the eigenvalues of the system and, thus, its analysis is interesting from the practical point of view. In this section, the eigenvalues are obtained by using the approach based on the Lambert W function and the results are discussed.

8.3.1   *Delay effects on rightmost eigenvalues*

Introducing a discrete delay in a system of DDEs changes the structure of the solution as seen in (2.43), which has the form of an infinite series with an infinite eigenspectrum. Figure 8.2 shows the change of the eigenspectrum by introducing a delay to the HIV model. If the system in Eq. (8.1) has no time-delay, all of the eigenvalues of the system are real (shown by the 'x' mark). However, as seen in Fig. 8.2, the time-delay leads to imaginary parts of the eigenvalues (shown by the 'o' mark) and, thus, to oscillations in the response. In the literature (Nelson *et al.*, 2000), time-delays, when used in population dynamic models, have been shown to create fluctuations in

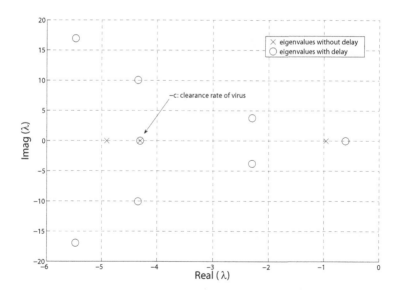

Fig. 8.2   Change of eigenvalues by introduction of delay in the HIV model: the rightmost eigenvalue shifted towards the imaginary axis. Also the time-delay leads to imaginary parts of the eigenvalues and, thus, to oscillations in the response.

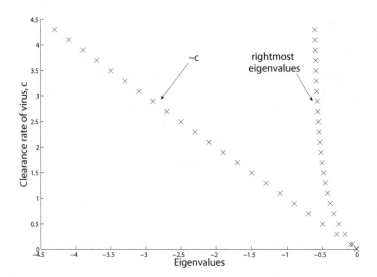

Fig. 8.3   As the clearance rate, c, declines, the rightmost eigenvalues, which are on the interval between $-c$ and the origin, moves toward to the right and, thus, the system becomes less stable.

population size. Without difficulty, from Eq. (8.3) it can be shown that one of the eigenvalues is always $-c$, regardless of the value of the time-delay, h, as seen in Fig. 8.2. Also, there exists one real eigenvalue of the system (8.1) on the interval between $-c$ and the origin (Nelson *et al.*, 2000). In the model of HIV, introduction of a time-delay makes the rightmost eigenvalues move to the right (i.e., less stable). This can be confirmed using an eigenvalue sensitivity analysis, which is introduced in the next section, as well as direct computation using the matrix Lambert W function as shown in Fig. 8.2. Because the eigenvalues of the HIV model describe the viral decline rate, via the eigenvalue changes shown in Fig. 8.2, it is confirmed that the delay reduces the long-term rate of decline of the viral load (Nelson *et al.*, 2000).

Also, depending on the parameters of the system, the stability can be determined via the rightmost eigenvalues. In Fig. 8.2, the system has one real rightmost eigenvalue of the system (8.1) on the interval between $-c$ and the origin. As the value of c declines, this rightmost eigenvalues moves toward to the origin and, thus, the system becomes less stable (see Fig. 8.3). This will be discussed more in detail via sensitivity analysis in the next section.

8.3.2 *Mutation, drug efficacy and eigenvalues*

Over the last decade, a number of potent drugs that inhibit HIV replication in vivo have been developed. Treatment regimes involving a combination of three or more different drugs can lead to a decline in viral load by several orders of magnitude. Although research is finding more drugs to combat HIV infection, the virus is continuously evolving to be resistant against these newly developed drugs. The high error rate in the reverse transcription process of viral RNA into DNA, combined with the continual viral replication of HIV, leads to the emergence of mutant strains of HIV that are drug resistant (D'Amato *et al.*, 2000). Most models for HIV assume either a perfect drug or an imperfect drug with a less than 100%, but constant, efficacy. In reality, the effect of antiviral treatment appears to change over time, due to i) pharmacokinetic variation, ii) fluctuating adherence, and iii) the emergence of drug resistant mutations (Huang *et al.*, 2003). Among them, drug resistance is a major concern in the treatment of some human infectious disease, especially, HIV. If strains that are resistant to the drug increase, then patients can become infected with the resistant virus, causing therapy to be ineffective (Wodarz and Lloyd, 2004). The result is a continuously varying efficacy of drug action. Accounting for this varying efficacy may be particularly important in recent clinical studies (Dixit and Perelson, 2004). The efficacy can be expressed as a function of time (see, e.g., (Huang *et al.*, 2003) and the references therein).

Although combination therapy can result in sustained suppression of viral load in many patients, it is not effective in all patients and fails after the emergence of drug-resistant strains. Hence, although finding new drugs to fight HIV is important for improving our chances for success, it is equally important to devise therapy regimes that minimize the chance of drug resistance emerging (Wodarz and Nowak, 2000). To do this, more detailed information about the status of patients and stability of the immune system of the patients with HIV is needed. Figure 8.4 shows the movement of the rightmost eigenvalue of the system with respect to drug efficacy. The rightmost eigenvalue moves toward the imaginary axis as the drug efficacy, n_p, decreases, and the status of the patient becomes less stable. This result tells us about the stability of the patient's immune system, and one can monitor the status of the immune system. Consequently, as time goes, the drug efficacy declines and the rightmost eigenvalue becomes larger and moves toward the imaginary axis (Fig. 8.4). Therefore, to sustain suppression of the viral load for AIDS patients, proactive switching and alternation of antiretroviral drug regimens is required (Martinez-Picado *et al.*, 2003).

Previously, the total viral load, $V_I + V_{NI}$, has been established as the primary prognostic indicator of progression to AIDS (D'Amato *et al.*, 2000), and the status of a patient's immune system is determined only in terms of viral load. However, the differences in parameters lead to widely varying conclusions about HIV pathogenesis (Ciupe *et al.*, 2006). Depending on the parameters involved in the system, such as δ and c, the viral load predictions can vary widely. Therefore, it would be more desirable to determine the stability of the immune system from the eigenvalues of the system, which is a function of the parameters involved in the model of HIV, in addition to the total viral load. Switching drugs too early risks poor adherence to a new drug regimen and may prematurely exhaust the limited number of remaining salvage therapies. Otherwise, switching too late leads to accumulation of mutations which leads to failure (i.e., viral rebound) (D'Amato *et al.*, 2000). The eigenvalue movement corresponding to the change of drug efficacy over time in Fig. 8.4 provides information about stability of the immune system for patients with HIV, and therapy regimes to sustain suppression of virus load continuously.

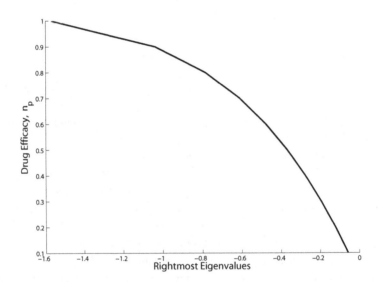

Fig. 8.4   Movement of the rightmost eigenvalues w.r.t drug efficacy: the rightmost eigenvalue moves to the imaginary axis as the drug efficacy, n_p decreases, and the status of the patient becomes less stable.

8.4   Sensitivity Analysis

In this section, eigenvalue and response sensitivity analysis with respect to parameters is considered. For systems of differential equations designed to model real systems, such as biological, chemical, or physical, one of the main goals is to understand the manner in which the parameters interact with properties of the systems, such as stability, dynamics, and response. These parameters are designed to correspond to aspects of the phenomena under investigation (e.g., productively infected T-cell death rate, δ, and clearance rate, c, in the HIV pathogenesis dynamics). Thus, it is desirable to predict how changes in the parameters will affect the system's properties: response and eigenvalues. Some previous work on this topic for the HIV model can be found in literature (see, e.g., (Banks and Bortz, 2005; Bortz and Nelson, 2004; Rong *et al.*, 2007), and the references therein).

Sensitivity analysis for eigenvalue and response, which has been developed in the context of modern control theory, can provide a mathematical tool for the model given by Eq. (8.1). The improved understanding the models can then help to design better experiments and develop better treatment regimes. Also, the interpretation of the results of sensitivity analysis for complex models makes it possible to understand which parameters have a greater influence on the response and/or eigenvalues. These parameters play an important role in the model and obtaining good estimates for them is critical especially when compared to other parameters to which solutions are less sensitive (Banks and Bortz, 2005).

8.4.1   *HIV: Eigenvalue sensitivity*

Sensitivity of the rightmost eigenvalues analysis reveals an understanding of the interactions of parameters with properties of systems, such as stability or movement behavior of state variables. Although a stability analysis was carried out using a random sampling method to identify which parameters are important in determining stability for systems of ODEs in (Rong *et al.*, 2007), the study on eigenvalue sensitivity analysis for an HIV model with a time-delay is presented here for the first time. The analytical expression for sensitivity of the rightmost eigenvalues can be derived by differentiating both sides of the characteristic Eq. (8.3) with respect to a parameter, say q, i.e.,

$$\frac{\partial H(\lambda)}{\partial q} = C(\lambda)\frac{\partial \lambda}{\partial q} + D(\lambda) = 0 \Rightarrow \frac{\partial \lambda}{\partial q} = -\frac{D(\lambda)}{C(\lambda)} \tag{8.4}$$

For example, the resulting sensitivity for the clearance rate of virus, c, the time-delay, h, productively infected T-cell death rate, δ, and for drug efficacy, n_p is given in Eqs. (8.5)–(8.8).

$$\frac{\partial\lambda}{\partial c} = -\frac{2\lambda^2 + (2\delta + 2c)\lambda + 2\delta c - (\lambda + c)(1 - n_p)\delta e^{-\lambda h}}{-(\lambda + c)(1 - n_p)\delta c e^{-\lambda h}(-h)} \cdots$$
$$\cdots \frac{-(1 - n_p)\delta c e^{-\lambda h}}{3\lambda^2 + 2(2c + \delta)\lambda + (2\delta c + c^2) - (1 - n_p)\delta c e^{-\lambda h}} \tag{8.5}$$

$$\frac{\partial\lambda}{\partial h} = \frac{(\lambda + c)\left\{\eta e^{-\lambda h}(-\lambda)\right\}}{3\lambda^2 + (2c + \delta)2\lambda + (2\delta c + c^2) + \eta e^{-\lambda h}(-1 + (\lambda + c)h)} \tag{8.6}$$

$$\frac{\partial\lambda}{\partial\delta} = -\frac{\lambda^2 + 2c\lambda + c^2}{3\lambda^2 + 2(2c + \delta)\lambda + (2\delta c + c^2) - (1 - n_p)\delta c e^{-\lambda h}} \cdots$$
$$\cdots \frac{-(\lambda + c)(1 - n_p)c e^{-\lambda h}}{-(\lambda + c)(1 - n_p)\delta c e^{-\lambda h}(-h)} \tag{8.7}$$

$$\frac{\partial\lambda}{\partial n_p} = -\frac{(\lambda + c)}{3\lambda^2 + 2(2c + \delta)\lambda + (2\delta c + c^2) - (1 - n_p)\delta c e^{-\lambda h} -} \cdots$$
$$\cdots \frac{\times \delta c e^{-\lambda h}}{(\lambda + c)(1 - n_p)\delta c e^{-\lambda h}(-h)} \tag{8.8}$$

With the parameter set in Table 8.1, the rightmost eigenvalues, λ_{rm}, of the system (8.1) from the previous section is (see Fig. 8.2)

$$\lambda_{rm} = -0.6118 \tag{8.9}$$

Then, by applying this rightmost eigenvalue and the parameter set in Table 8.1, the eigenvalue sensitivity is obtained from Eqs. (8.5)–(8.8) as

$$\frac{\partial\lambda_{rm}}{\partial h} = 0.4495, \quad \frac{\partial\lambda_{rm}}{\partial c} = -0.0173,$$
$$\frac{\partial\lambda_{rm}}{\partial\delta} = -0.1828, \quad \frac{\partial\lambda_{rm}}{\partial n_p} = -1.4983. \tag{8.10}$$

The signs determine whether a small increase in a parameter will increase or decrease the rightmost eigenvalue. If the sensitivity with respect to a parameter is positive, a small increase in the parameter makes the rightmost eigenvalues shift toward the right and, thus, the system becomes more unstable, and vice versa. As mentioned in the previous section (see Fig. 8.2), an increase of delay time destabilizes the system (sign of $\partial\lambda/\partial h$ is positive). For the other parameters, the sensitivities have negative signs, which means increases in the parameters stabilize the immune system and make the viral load decay faster. This can be inferred from the dynamics of Eq. (8.1): δ is the death rate of infected cells, which produce virus, and c is the clearance rate of virus. The sensitivity with respect to the clearance rate, c, of virus is relatively small, which means its effect on the stability of the immune system is not so significant compared to others. Also, because one infected T-cell produces N new virions, it can be inferred that the impact of variation in δ may have greater impact on the system than that of c, which explains why the magnitude of $\partial\lambda/\partial\delta$ is larger than that of $\partial\lambda/\partial c$. In this way, the impact of each parameter on the system is analyzed via the signs and the magnitudes. Note that this kind of analysis is possible only in linear cases, may not be feasible for nonlinear cases.

Also, as mentioned before, the parameters with high sensitivity should be given top priority when choosing which parameters to determine with a high degree of accuracy in estimating model parameters from data. To carry out parameter estimation for HIV models as in (8.1), one needs to specify a variance of each parameter in prior distribution (Huang *et al.*, 2003). Previously (e.g., in (Huang *et al.*, 2003; Wu *et al.*, 2005), etc.), if enough reliable information is available for some of the parameters, then small variances have been used, and vice versa. In such studies, the same variance has been given for c and δ, because enough prior information is available for both parameters. However, if sensitivity is analyzed as seen in (8.10), it is recommended to differentiate their variances more delicately depending on the sensitivity results, in order that a model may not be too sensitive to a specific parameter. By combining prior information and sensitivity analysis, more accurate estimation of parameters can be performed.

8.4.2   *Eigenvalue sensitivity and response sensitivity*

In (Bortz and Nelson, 2004), another type of sensitivity called response sensitivity was applied to the system (8.1). The response sensitivity analysis provides first-order estimates of the effect of parameter variations on the solutions. For the analysis, one needs to solve the state equation and a linear time-varying sensitivity equation simultaneously numerically, for example, using the delay differential equation solver *dde23* in Matlab (see Table 8.2). Considering the magnitudes and the signs, the result of a response sensitivity analysis presented in (Bortz and Nelson, 2004) shows good agreement with the eigenvalues sensitivities as in (8.10). The study

Table 8.2   Two different types of sensitivity for HIV model: eigenvalue sensitivity and response sensitivity (Yi *et al.*, 2008b).

	Response sensitivity	Eigenvalue sensitivity
For HIV Model	In (Bortz and Nelson, 2004)	In (Yi *et al.*, 2008b)
Objective	Effects of parameters on Response	Effects of parameters on Eigenvalues
Method	Numerical integration	Analytical derivation
Result Comparison	Show similar patterns in magnitudes and signs. See Eq. (8.13)	
Future Application	Useful in designing the optimal feedback control	Useful in designing the feedback control via eigenvalue assignment

in (Bortz and Nelson, 2004) showed that the response sensitivity with respect to the time-delay has a positive slope; on the other hand, the slopes of the response sensitivity with respect to c and δ are negative. Also, the absolute value of the response sensitivity with respect to c is smaller than that with respect to δ. Those coincide well with the results in (8.10). For rough comparison purposes, the response can be expressed in terms of the rightmost (i.e., dominant) eigenvalues and the initial condition as:

$$V(t) \approx e^{\lambda t} V_0 \qquad (8.11)$$

Taking derivatives of both sides with respect to a parameter yields

$$\underbrace{\frac{\partial V(t)}{\partial \rho}}_{\substack{\text{Response} \\ \text{Sensitivity}}} \approx e^{\lambda t} V_0 \underbrace{\frac{\partial \lambda}{\partial \rho}}_{\substack{\text{Eigenvalue} \\ \text{Sensitivity}}} t \qquad (8.12)$$

Then, Eq. (8.12) is divided by Eq. (8.11) to yield

$$\underbrace{\frac{\frac{\partial V(t)}{\partial \rho}}{V(t)}}_{\substack{\text{Normalized} \\ \text{Response} \\ \text{Sensitivity}}} \approx \underbrace{\frac{\partial \lambda}{\partial \rho}}_{\substack{\text{Eigenvalue} \\ \text{Sensitivity}}} \times t \qquad (8.13)$$

As seen in the approximation in Eq. (8.13), the normalized response sensitivity is proportional to the product of eigenvalue sensitivity and time. Even though a rough approximation, Eq. (8.13) can be helpful in conceptually relating the two different sensitivity approaches. Response sensitivity is a combined function of zero sensitivity and eigenvalue sensitivity (Rosenwasser and Yusupov, 2000). Therefore, for higher than first order systems of DDEs, it is not easy to derive an explicit relation between two sensitivities. However, Eq. (8.13) provides a good approximate relationship between them, based on the concept of dominant (i.e., rightmost) eigenvalues.

Using the eigenvalue sensitivities in (8.10), without integrating all state variables with respect to parameters of system as presented in (Bortz and Nelson, 2004), one can determine which parameter has the greatest influence on the system. This is achieved by comparing the magnitudes and signs of the eigenvalue sensitivity as in (8.10).

8.5 Concluding Remarks and Future Work

In this chapter, the model of HIV pathogenesis with an intracellular delay is considered. Because the model is represented by a system of DDEs, traditional approaches are not suitable for its analysis. Utilizing the Lambert W function-based approach developed in the previous chapters, the eigenvalues of the time-delay model of HIV pathogenesis are obtained. Furthermore, the approach is used to analyze changes in the eigenvalues of the HIV model as a time-delay is introduced. An increase in delay destabilizes the HIV system (see Fig. 8.2), and the result is confirmed via sensitivity analysis with respect to the time-delay, h (i.e., sign of $\partial \lambda / \partial h$ is positive). The movement of the rightmost eigenvalues with respect to the drug efficacy in the model is studied. Using the eigenvalues of the HIV model, the stability of the patients' immune system can be monitored. For example, corresponding to the change in drug efficacy due to mutation of the virus, the rightmost eigenvalues moves toward the right and the immune system of the patient becomes less stable (see Fig. 8.4). Because each patient has a different parameter set, the eigenvalues of the immune system can indicate progression to AIDS more accurately than just using total viral load, $V_I + V_{NI}$, as an indicator.

Sensitivity analysis was carried out with the rightmost eigenvalues obtained by using the Lambert W function. Sensitivities with respect to the parameters tell us about the impact of the variation of parameters on the immune system with HIV by their signs and magnitudes. For some parameters, the sensitivities have negative signs, which means an increase of the parameters stabilizes the immune system, and vice versa. Also, depending on the roles of the parameters, the magnitudes of sensitivities are different (e.g., c and δ). This sensitivity analysis can be used for various purposes, such as improved estimation of parameters, model validation, and design of therapy regimes by moving the rightmost eigenvalues by adjusting parameters, such as drug efficacy. Eigenvalues sensitivity with respect to each parameter of the system is expressed analytically in terms of the parameter, and shows good agreement with the response sensitivity result in (Bortz and Nelson, 2004). Unlike the response sensitivity approach, which integrates the sensitivity equation with a time-delay for a parameter set numerically over time, the eigenvalue sensitivity analysis is achieved analytically as in Eqs. (8.5)–(8.8).

Future work, based on the results presented in this chapter, may allow drug therapy design via the feedback control based on the Lambert W function (Yi *et al.*, 2010b) to be implemented with incomplete measurements and to minimize the expected effects of measurement error. For that, the controllability and observability analysis (Yi *et al.*, 2008a) for HIV model with a time-delay is also to be studied. Also, a similar approach can be applied to models for hepatitis B virus (HBV) infections (Ciupe *et al.*, 2007) and other viral dynamic models. One of the main goals for this research is to find more efficient and reliable therapy regimes.

Appendix A

Appendix for Chapter 2

A.1   Commutation of Matrices A and S in Eq. (2.10)

In general,

$$e^{\mathbf{X}}e^{\mathbf{Y}} = e^{\mathbf{X}+\mathbf{Y}} \tag{A.1}$$

is only true, when the matrices $\mathbf{X}$ and $\mathbf{Y}$ commute (i.e., $\mathbf{XY} = \mathbf{YX}$). As noted previously, the solution in Eq. (2.2) for the system in Eq. (2.1) is only valid when the matrices $\mathbf{S}$ and $\mathbf{A_d}$ commute, and in general they do not (see Eqs. (2.9) and (2.10)). Here it is shown that when the matrices $\mathbf{A_d}$ and $\mathbf{A}$ in Eq. (2.1) commute, then $\mathbf{S}$ and $\mathbf{A}$ will commute, and the solution in Eq. (2.2) becomes valid. From Eq. (2.14) it is noted that $\mathbf{S}$ can be expressed in terms of a polynomial function of the matrices $\mathbf{A_d}$ and $\mathbf{A}$, since both the exponential and Lambert W functions are represented as such polynomial series (Corless *et al.*, 1996). In general if two matrices $\mathbf{X}$ and $\mathbf{Y}$ commute, and the matrix functions $\mathbf{f}(\mathbf{X})$ and $\mathbf{g}(\mathbf{Y})$ can be expressed in a polynomial series form, i.e.,

$$\mathbf{f}(\mathbf{X}) = \sum_{k=k_0}^{k_1} p_k \mathbf{X}^k, \quad \mathbf{g}(\mathbf{Y}) = \sum_{k=k_0}^{k_1} q_k \mathbf{Y}^k \tag{A.2}$$

where p_k and q_k are arbitrary coefficients, then (Pease, 1965)

$$\mathbf{f}(\mathbf{X})\mathbf{g}(\mathbf{Y}) = \mathbf{g}(\mathbf{Y})\mathbf{f}(\mathbf{X}) \tag{A.3}$$

Consequently, if $\mathbf{A_d}$ and $\mathbf{A}$ commute, then $\mathbf{S}$ and $\mathbf{A}$ commute, and Eq. (2.2) is valid.

A.2   Reduction of Eqs. (2.31) and (2.32) to Eq. (2.36)

The Eq. (2.33) means that for $t \in [0, h]$,

$$\int_0^t e^{a(t-\xi)} bu(\xi)d\xi = \sum_{k=-\infty}^{\infty} \int_0^{t-h} e^{S_k(t-\xi)} C_k^N bu(\xi)d\xi + \int_{t-h}^t e^{a(t-\xi)} bu(\xi)d\xi$$

(A.4)

Continued from Eq. (A.4)

$$\int_0^t e^{a(t-\xi)} bu(\xi)d\xi - \int_{t-h}^t e^{a(t-\xi)} bu(\xi)d\xi = \sum_{k=-\infty}^{\infty} \int_0^{t-h} e^{S_k(t-\xi)} C_k^N bu(\xi)d\xi$$

$$\Rightarrow \int_0^{t-h} e^{a(t-\xi)} bu(\xi)d\xi = \int_0^{t-h} \sum_{k=-\infty}^{\infty} e^{S_k(t-\xi)} C_k^N bu(\xi)d\xi$$

$$\Rightarrow \int_0^{t-h} e^{a(t-\xi)} bu(\xi)d\xi - \int_0^{t-h} \sum_{k=-\infty}^{\infty} e^{S_k(t-\xi)} C_k^N bu(\xi)d\xi = 0$$

$$\Rightarrow \int_0^{t-h} \left\{ e^{a(t-\xi)} - \sum_{k=-\infty}^{\infty} e^{S_k(t-\xi)} C_k^N \right\} bu(\xi)d\xi = 0$$

$$\overset{(\theta=t-\xi)}{\Rightarrow} \int_t^h \left\{ e^{a\theta} - \sum_{k=-\infty}^{\infty} e^{S_k\theta} \right\} C_k^N bu(t-\theta)d\theta = 0, \quad \text{for } \forall t \in [0, h]$$

(A.5)

For the last equation in Eq. (A.5) to hold, for any value of $t \in [0, h]$, one can conclude as

$$\int_t^h \left\{ e^{a\theta} - \sum_{k=-\infty}^{\infty} e^{S_k\theta} C_k^N \right\} bu(t-\theta)d\theta = 0, \quad \text{for } \forall t \in [0, h]$$

(A.6)

$$\Rightarrow e^{a\theta} = \sum_{k=-\infty}^{\infty} e^{S_k\theta} C_k^N, \quad \text{where } \theta \in [0, h]$$

When the result in Eq. (A.6) is applied, Eqs. (2.31)–(2.32) can be combined into Eq. (2.36) after only some algebraic manipulation.

Appendix B

Appendix for Chapter 4

B.1 Proof Regarding Minimal Energy

Let us define

$$\bar{\mathbf{x}} = \mathbf{x}(t_1) - \mathbf{M}(t_1; 0, \mathbf{g}, \mathbf{x}_0) \tag{B.1}$$

Then the assumption that $\mathbf{u}'$ and $\mathbf{u}$ transfer $(\mathbf{x}_0, 0)$ to $(\mathbf{0}, t_1)$ implies that

$$\bar{\mathbf{x}} = \int_0^{t_1} \mathbf{K}(\xi, t_1)\mathbf{B}\mathbf{u}(\xi)d\xi = \int_0^{t_1} \mathbf{K}(\xi, t_1)\mathbf{B}\mathbf{u}'(\xi)d\xi \tag{B.2}$$

Subtracting both sides, one can obtain

$$\int_0^{t_1} \mathbf{K}(\xi, t_1)\mathbf{B}\{\mathbf{u}(\xi)' - \mathbf{u}(\xi)\}(\xi)d\xi = \mathbf{0} \tag{B.3}$$

which implies that

$$\left\langle \int_0^{t_1} \mathbf{K}(\xi, t_1)\mathbf{B}\{\mathbf{u}(\xi)' - \mathbf{u}(\xi)\}(\xi)d\xi, C_o^{-1}(0, t_1)\bar{\mathbf{x}} \right\rangle = 0 \tag{B.4}$$

where $< \, , \, >$ indicates the inner product of vectors. By using the following property of the inner product

$$\langle \mathbf{x}, \mathbf{A}\mathbf{y} \rangle = \left\langle \mathbf{A}^T\mathbf{x}, \mathbf{y} \right\rangle \tag{B.5}$$

this equation can rewritten as

$$\int_0^{t_1} \left\langle \mathbf{u}(\xi)' - \mathbf{u}(\xi), \{\mathbf{K}(\xi, t_1)\mathbf{B}\}^T C_o^{-1}(0, t_1)\bar{\mathbf{x}} \right\rangle d\xi = 0 \tag{B.6}$$

With the use of (4.5), and then (B.6) becomes

$$\int_0^{t_1} \langle \mathbf{u}(\xi)' - \mathbf{u}(\xi), \mathbf{u}(\xi) \rangle \, d\xi = 0 \tag{B.7}$$

Consider now

$$\int_0^{t_1} \|\mathbf{u}(\xi)'\|^2 \, d\xi \tag{B.8}$$

where $\|\mathbf{x}\| \equiv (\langle \mathbf{x}, \mathbf{x} \rangle)^{1/2}$. After some manipulation, and using (B.7), one can obtain

$$\begin{aligned}
\int_0^{t_1} \|\mathbf{u}(\xi)'\|^2 \, d\xi &= \int_0^{t_1} \|\mathbf{u}(\xi)' - \mathbf{u}(\xi) + \mathbf{u}(\xi)\|^2 \, d\xi \\
&= \int_0^{t_1} \|\mathbf{u}(\xi)' - \mathbf{u}(\xi)\|^2 \, d\xi + \int_0^{t_1} \|\mathbf{u}(\xi)\|^2 \, d\xi \\
&\quad + 2 \int_0^{t_1} \langle \mathbf{u}(\xi)' - \mathbf{u}(\xi), \mathbf{u}(\xi) \rangle \, d\xi \\
&= \int_0^{t_1} \|\mathbf{u}(\xi)' - \mathbf{u}(\xi)\|^2 \, d\xi + \int_0^{t_1} \|\mathbf{u}(\xi)\|^2 \, d\xi
\end{aligned} \tag{B.9}$$

Since

$$\int_0^{t_1} \|\mathbf{u}(\xi)' - \mathbf{u}(\xi)\|^2 \, d\xi \tag{B.10}$$

is always nonnegative, it can be concluded that

$$\int_0^{t_1} \|\mathbf{u}(\xi)'\|^2 \, d\xi \geq \int_0^{t_1} \|\mathbf{u}(\xi)\|^2 \, d\xi \tag{B.11}$$

$\square$

B.2   Comparisons with Other Types of Controllability and Observability

Depending on the nature of the problem under consideration, there exist various definitions of controllability and observability for time-delay systems. For example, spectral controllability has been developed to apply Finite Spectrum Assignment, a stabilizing method counteracting the effect of the delay based on prediction of the state. Spectral controllability is a sufficient condition for point-wise controllability used in our paper (sometimes point-wise controllability is also called controllability or fixed time complete controllability). The other definitions of controllability and observability are not related to *linear feedback controller* or *linear observer* as in systems of ODEs. The presented definitions and theorems in Chapter 4 are most similar to those for ODEs among the existing ones. The main purpose of the study in Chapter 4 is to put the controllability and observability Gramians to practical use by approximating them with the Lambert W function approach. Figures B.1 and B.2 show the relationships between various types of controllability and observability.

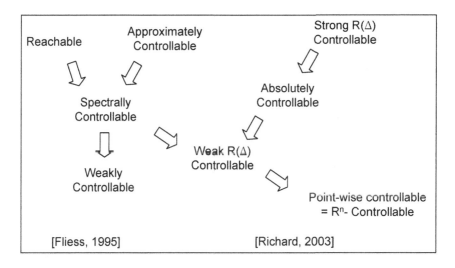

Fig. B.1   Relationship between various types of controllability.

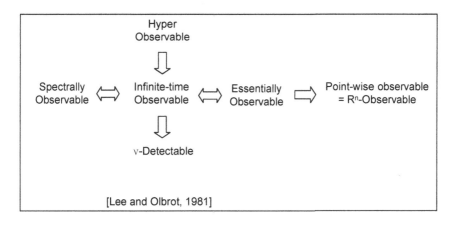

Fig. B.2   Relationship between various types of observability.

Appendix C

Appendix for Chapter 7

C.1 Limits in Assignment of Eigenvalues

One can choose feedback gains to assign the rightmost pole of systems with time-delays to desired positions in the complex plane based on the Lambert W function approach (Yi *et al.*, 2010b). Consider a simple, but unstable, example given by

$$\dot{x}(t) = ax(t) + u(t - h), \quad \text{where} \quad a > 0 \qquad (C.1)$$

with the linear feedback control

$$u(t) = fx(t) \qquad (C.2)$$

where f is a constant feedback gain. The time-delay in Eq. (C.1) can be caused by an inherent delay in the actuator. This formulation is also applicable to an open-loop system without delay but with a feedback delay. In these cases, the closed-loop system becomes

$$\dot{x}(t) = ax(t) + fx(t - h) \qquad (C.3)$$

The Lambert W function-based approach provides a method for design of feedback controllers via pole placement (Yi *et al.*, 2010b). The rightmost root of the characteristic equation of Eq. (C.3) is given by (Asl and Ulsoy, 2003)

$$S_0 = \frac{1}{h}W_0(fhe^{-ah}) + a \qquad (C.4)$$

The main goal is to choose the gain, f, to stabilize the system (C.1). One can stabilize the system (C.1) by setting the rightmost eigenvalue in Eq. (C.4), equal to the desired value:

$$\lambda_{des} = \frac{1}{h}W_0(fhe^{-ah}) + a \qquad (C.5)$$

Then, by solving Eq. (C.5), one can obtain an appropriate gain, f.

Example C.1. Assume $a = 1, h = 0.1$, and the desired rightmost eigenvalue, λ_{des}, is -1. Then, Eq. (C.5) is solved for the gain, f as

$$-1 = \frac{1}{0.1} W_0(f \times 0.1 \times e^{-0.1}) + 1 \tag{C.6}$$

Then, the solution, f is -1.8097.

Although a feedback controller can be designed for time-delay systems, the above result does not mean that the rightmost eigenvalue can be assigned arbitrarily. Depending on the parameters, especially the time-delay, of the system there exist limits and, in the worst case, some systems cannot be stabilized with any value of the feedback gain, f. As mentioned in Subsection 7.2.2, several different methods have been applied to investigate this problem. Here the problem is tackled by using the Lambert W function-based approach. As seen in Fig. C.1, each branch of the Lambert W function, $W_k(H)$ has its own range. Especially, for the principal branch, the real part of W_0 has a minimum value, -1, when H is $-1/e$ (Point A in Fig. C.2). Thus, the real part of W_0 is always equal to or larger than, -1, which leads to the following inequality regarding the lower limit in assigning the rightmost root,

$$\Re\{S_0\} = \Re\{\lambda_d\} = \frac{1}{h} \underbrace{\Re\{W_0(fhe^{-ah})\}}_{\geq -1} + a \geq -\frac{1}{h} + a \tag{C.7}$$

In this inequality the reciprocal of the time-delay, $1/h$, operates like a weighting factor. If h is smaller, the stabilizing term $W_0(fhe^{-ah})$, which can be adjusted with the feedback gain, f, has a relatively greater effect on S_0 and vice versa. Therefore, when the system is unstable (i.e., $a > 0$) it can be said that if the time delay h becomes larger, the system becomes more difficult to stabilize. Also, assuming that the feedback gain, f, can be selected to be any real value, even if the term, $W_0(fhe^{-ah})$ is minimized by choosing f as (see Fig. C.2)

$$fhe^{-ah} = -\frac{1}{e} \Rightarrow f = -\frac{1}{he^{1-ah}} \tag{C.8}$$

there exists a value, h^*, so that there is no possibility of stabilization for time delay, h, larger than h^*. That is,

$$0 = -\frac{1}{h^*} + a \Rightarrow h^* = \frac{1}{a} \tag{C.9}$$

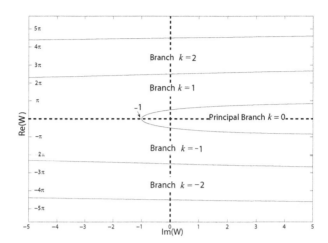

Fig. C.1 Ranges of each branch of the Lambert W function (Corless *et al.*, 1996). Real part of the principal branch, W_0, is equal or larger than -1 and this property sets limits on stabilization by feedback control.

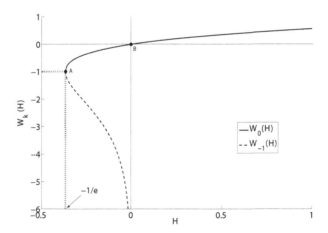

Fig. C.2 Real values of the branches, $k = 0$ (solid line) and $k = -1$ (dashed line), of the Lambert W function (Corless *et al.*, 1996).

For example, if $a = 1$, then for any $h > h^* = 1$ the system cannot be stabilized with any value of feedback gain, f. For instance, when $h = 2$ the rightmost eigenvalue, S_0, is always larger than $+0.5$ by the inequality (C.7). In this way, one can find the critical time delay for stabilization. If the feedback gain, f, has a specific limit on its value, for example, $f > 0$ in (C.7), then the argument of W_0 is also larger than zero. In that case, W_0 is also always larger than zero (see Fig. C.2), then the system can never be stabilized.

Also, depending on whether the time delay is in inputs, states and/or in the feedback control, there can exist many different cases for this problem. For example, one can consider a case other than (C.3), where:

$$\dot{x}(t) = ax(t) + a_d x(t - h) + u(t) \quad \text{where}$$
$$u(t) = fx(t) + f_d x(t - h) \tag{C.10}$$

Then, the rightmost root is given by

$$S_0 = \frac{1}{h} W_k \left((a_d + f_d) h e^{-(a+f)h} \right) + a + f \tag{C.11}$$

and the problems of stabilization, and its limitations, can be discussed in a similar way.

Example C.2. Assume $a = 1$, $a_d = 1$ (i.e., the system is unstable) in Eq. (C.10). Unlike the previous case, if there is no limits in choosing the gains, f and f_d, the system always can be stabilized independent of the value of h (e.g., when $f_d = -1$ and $f < -1$, the closed-loop system is stable). However, if $h = 2$ and f has to be larger than -0.5, then there is no way to stabilize the system with gains, f and f_d, which can be proven as in Eq. (C.7).

As mentioned previously, the same conclusion can be reached in different ways: using coordinate transformation in combination with a theorem in (Bellman and Cooke, 1963), or substituting a purely imaginary root in the characteristic equations to check bifurcation conditions (Beddington and May, 1975; Cooke and Grossman, 1982). Compared to such approaches, the approach using the Lambert W function, presented in this section, enables one to analyze the effect of delay directly from the solution form and to find these limits more intuitively.

Bibliography

Adams, B. M., Banks, H. T., Davidian, M., Kwon, H. D., Tran, H. T., Wynne, S. N. and Rosenberg, E. S. (2005). HIV dynamics: Modeling, data analysis, and optimal treatment protocols, *Journal of Computational and Applied Mathematics* **184**, 1, pp. 10–49.

Adams, B. M., Banks, H. T., Kwon, H. D. and Tran, H. T. (2004). Dynamic multidrug therapies for HIV: Optimal and STI control approaches, *Mathematical Biosciences and Engineering* **1**, 2, pp. 223–241.

Asl, F. M. and Ulsoy, A. G. (2003). Analysis of a system of linear delay differential equations, *Journal of Dynamic Systems, Measurement and Control* **125**, 2, pp. 215–223.

Asl, F. M. and Ulsoy, A. G. (2007). Closure to "Discussion of 'analysis of a system of linear delay differential equations' ", *Journal of Dynamic Systems Measurement and Control* **129**, 1, pp. 123–123.

Banks, H. T. and Bortz, D. M. (2005). A parameter sensitivity methodology in the context of hiv delay equation models, *Journal of Mathematical Biology* **50**, pp. 607–625.

Banks, H. T. and Manitius, A. (1975). Projection series for retarded functional differential equations with applications to optimal control problems, *Journal of Differential Equatios* **18**, 2, pp. 296–332.

Beddington, J. R. and May, R. M. (1975). Time delays are not necessarily destabilizing, *Mathematical Biosciences* **27**, 1-2, pp. 109–117.

Bellman, R. E. and Cooke, K. L. (1963). *Differential-Difference Equations* (Academic Press, New York).

Bengea, S. C., Li, X. Q. and DeCarlo, R. A. (2004). Combined controller-observer design for uncertain time delay systems with application to engine idle speed control, *Journal of Dynamic Systems Measurement and Control* **126**, 4, pp. 772–780.

Bhat, K. P. M. and Koivo, H. N. (1976a). Modal characterizations of controllability and observability in time delay systems, *IEEE Transactions on Automatic Control* **21**, 2, pp. 292–293.

Bhat, K. P. M. and Koivo, H. N. (1976b). An observer theory for time delay systems, *IEEE Transactions on Automatic Control* **21**, 2, pp. 266–269.

Bortz, D. M. and Nelson, P. W. (2004). Sensitivity analysis of a nonlinear lumped parameter model of HIV infection dynamics, *Bulletin of Mathematical Biology* **66**, 5, pp. 1009–1026.

Bortz, D. M. and Nelson, P. W. (2006). Model selection and mixed-effects modeling of HIV infection dynamics, *Bulletin of Mathematical Biology* **68**, 8, pp. 2005–2025.

Brethe, D. and Loiseau, J. J. (1998). An effective algorithm for finite spectrum assignment of single-input systems with delays, *Mathematics and Computers in Simulation* **45**, 3-4, pp. 339–348.

Buckalo, A. F. (1968). Explicit conditions for controllability of linear systems with time lag, *IEEE Transactions on Automatic Control* **13**, pp. 193–195.

Chen, C. T. (1984). *Linear System Theory and Design* (Holt, Rinehart and Winston, New York).

Chen, S. G., Ulsoy, A. G. and Koren, Y. (1997). Computational stability analysis of chatter in turning, *Journal of Dynamic Systems, Measurement and Control* **119**, 4, pp. 457–460.

Chen, Y. and Moore, K. L. (2002a). Analytical stability bound for a class of delayed fractional-order dynamic systems, *Nonlinear Dynamics* **29**, 1-4, pp. 191–200.

Chen, Y. and Moore, K. L. (2002b). Analytical stability bound for delayed second-order systems with repeating poles using Lambert function W, *Automatica* **38**, 5, pp. 891–895.

Cheng, Y. and Hwang, C. (2006). Use of the Lambert W function for time-domain analysis of feedback fractional delay systems, *IEE Proceedings: Control Theory and Applications* **153**, 2, pp. 167–174.

Choudhury, A. K. (1972a). Algebraic and transfer-function criteria of fixed-time controllability of delay-differential systems, *International Journal of Control* **16**, 6, pp. 1073–1082.

Choudhury, A. K. (1972b). Necessary and sufficient conditions of pointwise completeness of linear time-invariant delay-differential systems, *Internation Journal of Control* **16**, 6, pp. 1083–1100.

Ciupe, M. S., Bivort, B. L., Bortz, D. M. and Nelson, P. W. (2006). Estimating kinetic parameters from HIV primary infection data through the eyes of three different mathematical models, *Mathematical Biosciences* **200**, 1, pp. 1–27.

Ciupe, S. M., Ribeiro, R. M., Nelson, P. W., Dusheiko, G. and Perelson, A. S. (2007). The role of cells refractory to productive infection in acute hepatitis B viral dynamics, *Proceedings of The National Academy of Sciences of the United States of America* **104**, 12, pp. 5050–5055.

Cooke, K. L. and Grossman, Z. (1982). Discrete delay, distributed delay and stability switches, *Journal of Mathematical Analysis and Applications* **86**, 2, pp. 592–627.

Corless, R. M., Gonnet, G. H., Hare, D. E. G., Jeffrey, D. J. and Knuth, D. E. (1996). On the Lambert W function, *Advances in Computational Mathematics* **5**, 4, pp. 329–359.

D'Amato, R. M., D'Aquila, R. T. and Wein, L. M. (2000). Management of antiretroviral therapy for HIV infection: Analyzing when to change therapy, *Management Science* **46**, 9, pp. 1200–1213.

Darouach, M. (2001). Linear functional observers for systems with delays in state variables, *IEEE Transactions on Automatic Control* **46**, 3, pp. 491–496.

de la Sen, M. (2005). On pole-placement controllers for linear time-delay systems with commensurate point delays, *Mathematical Problems in Engineering*, 1, pp. 123–140.

Delfour, M. C. and Mitter, S. K. (1972). Controllability, observability and optimal feedback control of affine hereditary differential systems, *SIAM Journal of Control* **10**, pp. 298–328.

Dixit, N. M. and Perelson, A. S. (2004). Complex patterns of viral load decay under antiretroviral therapy: influence of pharmacokinetics and intracellular delay, *Journal of Theoretical Biology* **226**, 1, pp. 95–109.

Engelborghs, K., Dambrine, M. and Roose, D. (2001). Limitations of a class of stabilization methods for delay systems, *IEEE Transactions on Automatic Control* **46**, 2, pp. 336–339.

Fattouh, A., Sename, O. and Dion, J. M. (1998). H-infinity observer design for time-delay systems, in *Proceedings of the 37th IEEE Conference on Decision and Control, Tampa FL, Dec. 1998*, pp. 4545–4546.

Fliess, M., Marquez, R. and Mounier, H. (2002). An extension of predictive control, PID regulators and smith predictors to some linear delay systems, *International Journal of Control* **75**, 10, pp. 728–743.

Fofana, M. S. (2003). Delay dynamical systems and applications to nonlinear machine-tool chatter, *Chaos Solitons & Fractals* **17**, pp. 731–747.

Forde, J. and Nelson, P. (2004). Applications of sturm sequences to bifurcation analysis of delay differential equation models, *Journal of Mathematical Analysis and Applications* **300**, 2, pp. 273–284.

Franklin, G. F., Powell, J. D. and Emami-Naeini, A. (2005). *Feedback Control of Dynamic Systems* (Pearson Prentice Hall, Upper Saddle River, NJ).

Frost, M. G. (1982). Controllability, observability and the transfer function matrix for a delay-differential system, *Internation Journal of Control* **35**, 1, pp. 175–182.

Gabasov, R., Zhevnyak, R. M., Kirillova, F. M. and Kopeikina, T. B. (1972). Relative observability of linear systems. 2. observations under constantly acting perturbations, *Automation and Remote Control* **33**, 10, pp. 1563–1573.

Glizer, V. Y. (2004). Observability of singularly perturbed linear time-dependent differential systems with small delay, *Journal of Dynamical and Control Systems* **10**, 3, pp. 329–363.

Gorecki, H., Fuksa, S., Grabowski, P. and Korytowski, A. (1989). *Analysis and Synthesis of Time Delay Systems* (John Wiley & Sons, New York).

Gu, K. and Niculescu, S. I. (2006). Stability analysis of time-delay systems: A Lyapunov approach, *Advanced Topics in Control Systems Theory, Lecture Notes in Control and Information Sciences* **328**, pp. 139–170.

Hale, J. K. and Lunel, S. M. V. (1993). *Introduction to Functional Differential Equations* (Springer-Verlag, New York).

Herz, A. V. M., Bonhoeffer, S., Anderson, R. M., May, R. M. and Nowak, M. A. (1996). Viral dynamics in vivo: Limitations on estimates of intracellular delay and virus decay, *Proceedings of the National Academy of Sciences of The United States of America* **93**, 14, pp. 7247–7251.

Ho, D. D., Neumann, A. U., Perelson, A. S., Chen, W., Leonard, J. M. and Markowitz, M. (1995). Rapid turnover of plasma virions and CD4 lymphocites in HIV-1 infection, *Nature* **373**, 6510, pp. 123–126.

Holford, S. and Agathoklis, P. (1996). Use of model reduction techniques for designing IIR filters with linear phase in the passband, *IEEE Transactions on Signal Processing* **44**, 10, pp. 2396–2404.

Hovel, P. and Scholl, E. (2005). Control of unstable steady states by time-delayed feedback methods, *Physical Review E* **72**, 4, pp. 1–7.

Hrissagis, K. and Kosmidou, O. I. (1998). Delay-dependent robust stability conditions and decay estimates for systems with input delays, *Kybernetika* **34**, 6, pp. 681–691.

Hu, G. and Davison, E. J. (2003). Real stability radii of linear time-invariant time-delay systems, *Systems and Control Letters* **50**, 3, pp. 209–219.

Huang, Y. X., Rosenkranz, S. L. and Wu, H. L. (2003). Modeling HIV dynamics and antiviral response with consideration of time-varying drug exposures, adherence and phenotypic sensitivity, *Mathematical Biosciences* **184**, 2, pp. 165–186.

Hwang, C. and Cheng, Y. (2005). A note on the use of the Lambert W function in the stability analysis of time-delay systems, *Automatica* **41**, 11, pp. 1979–1985.

Jankovic, M. (2001). Control design for a diesel engine model with time delay, in *Proceedings of the 40th IEEE Conference on Decision and Control, Orlando, FL, Dec. 2001*, pp. 117–122.

Jankovic, M. and Kolmanovsky, I. (2000). Constructive Lyapunov control design for turbocharged diesel engines, *IEEE Transactions on Control Systems Technology* **8**, 2, pp. 288–299.

Jankovic, M. and Kolmanovsky, I. (2009). *Developments in Control of Time-Delay Systems for Automotive Powertrain Applications, Delay Differential Equations: Recent Advances and New Directions* (Springer).

Jarlebring, E. and Damm, T. (2007). The Lambert W function and the spectrum of some multidimensional time-delay systems, *Automatica* **43**, 12, pp. 2124–2128.

Kalmar-Nagy, T., Stepan, G. and Moon, F. C. (2001). Subcritical hopf bifurcation in the delay equation model for machine tool vibrations, *Nonlinear Dynamics* **26**, pp. 121–142.

Kawabata, K. and Mori, T. (2009). Feedback enlargement of stability radius by nondifferentiable optimization, *Electrical Engineering in Japan* **166**, 3, pp. 55–61.

Kaya, I. (2004). IMC based automatic tuning method for PID controllers in a Smith predictor configuration, *Computers & Chemical Engineering* **28**, 3, pp. 281–290.

Kirillova, F. M. and Churakova, S. V. (1967). The controllability problem for linear systems with aftereffect, *Differential Equations* **3**, pp. 221–225.

Kirschner, D. and Webb, G. F. (1996). A model for treatment strategy in the chemotherapy of AIDS, *Bulletin of Mathematical Biology* **58**, 2, pp. 367–390.

Kopeikina, T. B. (1998). Relative observability of linear nonstationary singularly perturbed delay systems, *Differential Equations* **34**, 1, pp. 22–28.

Kuang, Y. (1993). *Delay Differential Equations: with Applications in Population Dynamics* (Academic Press, Cambridge, Mass.).

Lee, E. B. and Olbrot, A. (1981). Observability and related structural results for linear hereditary systems, *International Journal of Control* **34**, 6, pp. 1061–1078.

Lee, T. N. and Dianat, S. (1981). Stability of time-delay systems, *IEEE Transactions on Automatic Control* **26**, 4, pp. 951–953.

Leyvaramos, J. and Pearson, A. E. (1995). An asymptotic modal observer for linear autonomous time-lag systems, *IEEE Transactions on Automatic Control* **40**, 7, pp. 1291–1294.

Li, X. and deSouza, C. E. (1998). Output feedback stabilization of linear time-delay systems, in L. Dugard and E. I. Verriest (eds.), *Stability and Control of Time-delay Systems* (Springer, New York), pp. 241–258.

Li, X., Ji, J. C., Hansen, C. H. and Tan, C. (2006). The response of a Duffing-van der Pol oscillator under delayed feedback control, *Journal of Sound and Vibration* **291**, 3-5, pp. 644–655.

Lien, C. H., Sun, Y. J. and Hsieh, J. G. (1999). Global stabilizability for a class of uncertain systems with multiple time-varying delays via linear control, *International Journal of Control* **72**, 10, pp. 904–910.

Liu, P. L. (2003). Exponential stability for linear time-delay systems with delay dependence, *Journal of the Franklin Institute* **340**, 6-7, pp. 481–488.

Lunel, S. M. V. (1989). *Exponential Type Calculus for Linear Delay Equations* (Centrum voor Wiskunde en Informatica, Amsterdam, The Netherlands).

Maccari, A. (2001). The response of a parametrically excited van der Pol oscillator to a time delay state feedback, *Nonlinear Dynamics* **26**, 2, pp. 105–119.

Maccari, A. (2003). Vibration control for the primary resonance of the van der Pol oscillator by a time delay state feedback, *International Journal of Non-Linear Mechanics* **38**, 1, pp. 123–131.

Mahmoud, M. S. (2000). *Robust Control and Filtering for Time-Delay Systems* (Marcel Dekker, New York).

Mahmoud, M. S. and Ismail, A. (2005). New results on delay-dependent control of time-delay systems, *IEEE Transactions on Automatic Control* **50**, 1, pp. 95–100.

Malek-Zavarei, M. and Jamshidi, M. (1987). *Time-delay Systems: Analysis, Optimization, and Applications* (Elsevier Science Pub., New York, U.S.A.).

Manitius, A. and Olbrot, A. W. (1979). Finite spectrum assignment problem for systems with delays, *IEEE Transactions on Automatic Control* **24**, 4, pp. 541–553.

Mao, W. J. and Chu, J. (2006). D-stability for linear continuous-time systems with multiple time delays, *Automatica* **42**, 9, pp. 1589–1592.

Marino, S., Beretta, E. and Kirschner, D. E. (2007). The role of delays in innate and adaptive immunity to intracellular bacterial infection, *Mathematical Biosciences and Engineering* **4**, 2, pp. 261–286.

Martinez-Picado, J., Negredo, E., Ruiz, L., Shintani, A., Fumaz, C. R., Zala, C., Domingo, P., Vilaro, J., Llibre, J. M., Viciana, P., Hertogs, K., Boucher, C., D'Aquila, R. T. and Clotet, B. (2003). Alternation of antiretroviral drug regimens for HIV infection - a randomized, controlled trial, *Annals of Internal Medicine* **139**, 2, pp. 81–89.

Merritt, H. E. (1965). Theory of self-excited machine-tool chatter-contribution to machine-too chatter research-1, *Journal of Engineering for Industry* **87**, 4, pp. 447–454.

Michiels, W., Engelborghs, K., Vansevenant, P. and Roose, D. (2002). Continuous pole placement for delay equations, *Automatica* **38**, 5, pp. 747–761.

Michiels, W. and Roose, D. (2003). An eigenvalue based approach for the robust stabilization of linear time-delay systems, *International Journal of Control* **76**, 7, pp. 678–686.

Minis, I. E., Magrab, E. B. and Pandelidis, I. O. (1990). Improved methods for the prediction of chatter in turning, part 3. a generalized linear theory, *Journal of Engineering for Industry* **112**, 1, pp. 28–35.

Mittler, J. E., Sulzer, B., Neumann, A. U. and Perelson, A. S. (1998). Influence of delayed viral production on viral dynamics in HIV-1 infected patients, *Mathematical Biosciences* **152**, 2, pp. 143–163.

Moore, B. C. (1981). Principal component analysis in linear systems: controllability, observability, and model reduction, *IEEE Transactions on Automatic Control* **26**, 1, pp. 17–32.

Mounier, H. (1998). Algebraic interpretation of the spectral controllability of a linear delay system, *Forum Mathematicum* **10**, 1, pp. 39–58.

Nayfeh, A., Chin, C., Pratt, J. and Moon, F. C. (1997). Application of perturbation methods to tool chatter dynamics, in F. C. Moon (ed.), *Dynamics and Chaos in Manufacturing Processes* (Wiley, New York), pp. 193–213.

Nelson, P. W., Mittler, J. E. and Perelson, A. S. (2001). Effect of drug efficacy and the eclipse phase of the viral life cycle on estimates of HIV viral dynamic parameters, *Journal of Acquired Immune Deficiency Syndromes* **26**, 5, pp. 405–412.

Nelson, P. W., Murray, J. D. and Perelson, A. S. (2000). A model of HIV-1 pathogenesis that includes an intracellular delay, *Mathematical Biosciences* **163**, 2, pp. 201–215.

Nelson, P. W. and Perelson, A. S. (2002). Mathematical analysis of delay differential equation models of HIV-1 infection, *Mathematical Biosciences* **179**, 1, pp. 73–94.

Niculescu, S. I. (1998). H-infinity memoryless control with an alpha-stability constraint for time-delay systems: An LMI approach, *IEEE Transactions on Automatic Control* **43**, 5, pp. 739–743.

Niculescu, S. I. (2001). *Delay Effects on Stability: a Robust Control Approach* (Springer, New York).

Niculescu, S. I. and Annaswamy, A. M. (2003). An adaptive Smith-controller for time-delay systems with relative degree $n^* <= 2$, *Systems & Control Letters* **49**, 5, pp. 347–358.

Nowak, M. A., Bonhoeffer, S., Shaw, G. M. and May, R. M. (1997). Anti-viral drug treatment: Dynamics of resistance in free virus and infected cell populations, *Journal of Theoretical Biology* **184**, 2, pp. 205–219.

Olbrot, A. W. (1972). On controllability of linear systems with time delays in control, *IEEE Transactions on Automatic Control* **17**, 5, pp. 664–666.

Olgac, N., Elmali, H., Hosek, M. and Renzulli, M. (1997). Active vibration control of distributed systems using delayed resonator with acceleration feedback, *Journal of Dynamic Systems, Measurement and Control* **119**, 3, pp. 380–389.

Olgac, N., Ergenc, A. and Sipahi, R. (2005). "Delay scheduling": A new concept for stabilization in multiple delay systems, *Journal of Vibration and Control* **11**, 9, pp. 1159–1172.

Olgac, N. and Sipahi, R. (2002). An exact method for the stability analysis of time-delayed linear time-invariant (LTI) systems, *IEEE Transactions on Automatic Control* **47**, 5, pp. 793–797.

Olgac, N. and Sipahi, R. (2005). A unique methodology for chatter stability mapping in simultaneous machining, *Journal of Manufacturing Science and Engineering* **127**, 4, pp. 791–800.

Optiz, H. and Bernardi, F. (1970). Investigation and calculation of the chatter behavior of lathes and milling machines, *Annlas of CIRP* **18**, pp. 335–343.

Pearson, A. E. and Fiagbedzi, Y. A. (1989). An obsever for time lag systems, *IEEE Transactions on Automatic Control* **34**, 7, pp. 775–777.

Pease, M. C. (1965). *Methods of Matrix Algebra* (Academic Press, New York).

Perelson, A. S. (2002). Modelling viral and immune system dynamics, *Nature Reviews Immunology* **2**, 1.

Perelson, A. S. and Nelson, P. W. (1999). Mathematical analysis of HIV-1 dynamics in VIVO, *SIAM Review* **41**, 1, pp. 3–44.

Perelson, A. S., Neumann, A. U., Markowitz, M., Leonard, J. M. and Ho, D. D. (1996). HIV-1 dynamics in vivo: Virion clearance rate, infected cell lifespan, and viral generation time, *Science* **271**, 5255, pp. 1582–1586.

Pila, A. W., Shaked, U. and de Souza, C. E. (1999). H-infinity filtering for continuous-time linear systems with delay, *IEEE Transactions on Automatic Control* **44**, 7, pp. 1412–1417.

Postlethwaite, I. and Foo, Y. K. (1985). Robustness with simultaneous pole and zero movement across the j-omega-axis, *Automatica* **21**, 4, pp. 433–443.

Pyragas, K. (1992). Continuous control of chaos by self-controlling feedback, *Physics Letters A* **170**, 6, pp. 421–428.

Qiu, L., Bernhardsson, B., Rantzer, A., Davison, E. J., Young, P. M. and Doyle, J. C. (1995). A formula for computation of the real stability radius, *Automatica* **31**, 6, pp. 879–890.

Radjavi, H. and Rosenthal, P. (2000). *Simultaneous Triangularization* (Springer, New York).

Reddy, D. V. R., Sen, A. and Johnston, G. L. (2000). Dynamics of a limit cycle oscillator under time delayed linear and nonlinear feedbacks, *Physica D: Nonlinear Phenomena* **144**, 3, pp. 335–357.

Richard, J. P. (2003). Time-delay systems: an overview of some recent advances and open problems, *Automatica* **39**, 10, pp. 1667–1694.

Rong, L. B., Feng, Z. L. and Perelson, A. S. (2007). Emergence of HIV-1 drug resistance during antiretroviral treatment, *Bulletin of Mathematical Biology* **69**, pp. 2027–2060.

Rosenwasser, Y. and Yusupov, R. (2000). *Sensitivity of Automatic Control Systems* (CRC Press, Boca Raton, Fla.).

Salamon, D. (1980). Observers and duality between observation and state feedback for time-delay systems, *IEEE Transactions on Automatic Control* **25**, 6, pp. 1187–1192.

Shafiei, Z. and Shenton, A. T. (1997). Frequency-domain design of PID controllers for stable and unstable systems with time delay, *Automatica* **33**, 12, pp. 2223–2232.

Sharifi, F. J., Beheshti, M. T. H., Ganjefar, S. and Momeni, H. (2003). Behavior of Smith predictor in teleoperation systems with modeling and delay time errors, in *Proceedings of 2003 IEEE Conference on Control Applications, Maui, HI, Dec. 2003*, pp. 1176–1180.

Shinozaki, H. and Mori, T. (2006). Robust stability analysis of linear time-delay systems by Lambert W function: Some extreme point results, *Automatica* **42**, 10, pp. 1791–1799.

Silva, G. J. and Datta, A. B. S. P. (2001). Controller design via Padé approximation can lead to instability, in *Proceedings of the 40th IEEE Conference on Decision and Control, Orlando, FL, Dec. 2001*, pp. 4733–4737.

Sipahi, R. and Olgac, N. (2003a). Active vibration suppression with time delayed feedback, *Journal of Vibration and Acoustics* **125**, 3, pp. 384–388.

Sipahi, R. and Olgac, N. (2003b). Degenerate cases in using the direct method, *Journal of Dynamic Systems Measurement and Control* **125**, 2, pp. 194–201.

Smith, O. (1957). Closer control of loops with dead time, *Chemical Engineering Progress* **53**, pp. 217–219.

Stepan, G. (1989). *Retarded Dynamical Systems: Stability and Characteristic Functions* (Wiley, New York).

Stepan, G. and Insperger, T. (2006). Stability of time-periodic and delayed systems - a route to act-and-wait control, *Annual reviews in control* **30**, 2, pp. 159–168.

Stepan, G. and Moon, F. C. (1997). Delay differential equation models for machine tool chatter, in F. C. Moon (ed.), *Dynamics and Chaos in Manufacturing Processes* (Wiley, New York), pp. 165–191.

Suh, B. S. and Yang, J. H. (2005). A tuning of PID regulators via LQR approach, *Journal of Chemical Engineering of Japan* **38**, 5, pp. 344–356.

Thowsen, A. (1977). On pointwise degeneracy, controllability and minimal time control of linear dynamical systems with delays, *International Journal of Control* **25**, 3, pp. 345–360.

Tlusty, J. (2000). *Manufacturing Processes and Equipment* (Prentice Hall, New Jersey).

Tobias, S. A. (1965). *Machine-tool Vibration* (Wiley, New York).

Trinh, H. (1999). Linear functional state observer for time-delay systems, *International Journal of Control* **72**, 18, pp. 1642–1658.

Tsoi, A. C. and Gregson, M. J. (1978). Recent advances in the algebraic system theory of delay differential equations, in M. J. Gregson (ed.), *Recent Theoretical Developments in Control* (Academic Press, New York), pp. 67–127.

Van Assche, V., Dambrine, M., Lafay, J. F. and Richard, J. P. (1999). Some problems arising in the implementation of distributed-delay control laws, in *Proceedings of 38th IEEE Conference on Decision and Control, Phoenix, AZ, Dec. 1999*, pp. 4668–4672.

Vandevenne, H. F. (1972). Controllability and stabilizability properties of delay systems, in *Proceedings of the 1972 IEEE Conference on Decision Control and 11th Symposium on Adaptive Processes, Phoenix, AZ, Dec. 1999*, pp. 370–377.

Verriest, E. I. and Kailath, T. (1983). On generalized balanced realizations, *IEEE Transcations on Automatic Control* **28**, 8, pp. 833–844.

Wang, Q., Lee, H. T. and Tan, K. K. (1995). Automatic tuning of finite spectrum assignment controllers for delay systems, *Automatica* **31**, 3, pp. 477–482.

Wang, Q. G., Liu, M. and Hang, C. C. (2007). Approximate pole placement with dominance for continuous delay systems by PID controllers, *Canadian Journal of Chemical Engineering* **85**, 4, pp. 549–557.

Wang, Z. H. and Hu, H. Y. (2007). Robust stability of time-delay systems with uncertain parameters, *IUTAM Symposium on Dynamics and Control of Nonlinear Systems with Uncertainty* **2**, pp. 363–372.

Wang, Z. H. and Hu, H. Y. (2008). Calculation of the rightmost characteristic root of retarded time-delay systems via Lambert W function, *Journal of Sound and Vibration* **318**, 4-5, pp. 757–767.

Weiss, L. (1967). On the controllability of delay-differential equations, *SIAM Journal on Control and Optimization* **5**, 4, pp. 575–587.

Weiss, L. (1970). An algebraic criterion for controllability of linear systems with time delay, *IEEE Transactions on Automatic Control* **15**, 4, pp. 443–444.

Wirkus, S. and Rand, R. (2002). The dynamics of two coupled van der pol oscillators with delay coupling, *Nonlinear Dynamics* **30**, 3, pp. 205–221.

Wodarz, D. and Lloyd, A. L. (2004). Immune responses and the emergence of drug-resistant virus strains in vivo, *Proceedings of the Royal Society of London Series B-Biological Sciences* **271**, pp. 1101–1109.

Wodarz, D. and Nowak, M. A. (2000). HIV therapy: Managing resistance, *Proceedings of the National Academy of Sciences of the United States of America* **97**, 15, pp. 8193–8195.

Wu, H. L., Huang, Y. X., Acosta, P., Rosenkranz, S. L., Kuritzkes, D. R., Eron, J. J., Perelson, A. S. and Gerber, J. G. (2005). Modeling long-term HIV dynamics and antiretroviral response - effects of drug potency, pharmacokinetics, adherence, and drug resistance, *Journal of Acquired Immune Deficiency Syndromes* **39**, 3, pp. 272–283.

Xu, J. and Chung, K. W. (2003). Effects of time delayed position feedback on a van der pol-duffing oscillator, *Physica D: Nonlinear Phenomena* **180**, 1-2, pp. 17–39.

Yi, S., Nelson, P. W. and Ulsoy, A. G. (2006a). Chatter stability analysis using the matrix lambert function and bifurcation analysis, in *Proceedings of the 2006 ASME International Conference on Manufacturing Science and Engineering, Ypsilanti, MI, Sept. 2006*, MSEC 2006-21131.

Yi, S., Nelson, P. W. and Ulsoy, A. G. (2007a). Controllability and observability of systems of linear delay differential equations via the matrix Lambert W function, in *Proceedings of 2007 American Control Conference, New York, NY, July 2007*, pp. 5631–5636.

Yi, S., Nelson, P. W. and Ulsoy, A. G. (2007b). Delay differential equations via the matrix Lambert W function and bifurcation analysis: Application to machine tool chatter, *Mathematical Biosciences and Engineering* **4**, 2, pp. 355–368.

Yi, S., Nelson, P. W. and Ulsoy, A. G. (2007c). Feedback control via eigenvalue assignment for time delayed systems using the Lambert W function, in *Proceedings of the ASME 2007 IDETC, Las Vegas, NV, Sept. 2007*, DETC 2007-35711.

Yi, S., Nelson, P. W. and Ulsoy, A. G. (2007d). Survey on analysis of time delayed systems via the Lambert W function, *Dynamics of Continuous, Discrete and Impulsive Systems (Series A)* **14**, S2, pp. 296–301.

Yi, S., Nelson, P. W. and Ulsoy, A. G. (2008a). Controllability and observability of systems of linear delay differential equations via the matrix Lambert W function, *IEEE Transactions on Automatic Control* **53**, 3, pp. 854–860.

Yi, S., Nelson, P. W. and Ulsoy, A. G. (2008b). Eigenvalues and sensitivity analysis for a model of HIV-1 pathogenesis with an intracellular delay, in *Proceedings of 2008 ASME Dynamic Systems and Control Conference, Ann Arbor, MI, Oct. 2008*, DSCC2008-2408.

Yi, S., Nelson, P. W. and Ulsoy, A. G. (2008c). Robust control and time-domain specifications for systems for delay differential equations via eigenvalue assignment, in *Proceedings of 2008 American Control Conference, Seattle, WA, June 2008*, pp. 4298–4933.

Yi, S., Nelson, P. W. and Ulsoy, A. G. (2009). Design of observer-based feedback control for time-delay systems with application to automotive powertrain control, in *Proceedings of 2009 ASME Dynamic Systems and Control Conference, Hollywood, CA, Oct. 2009*, DSCC2009-2590.

Yi, S., Nelson, P. W. and Ulsoy, A. G. (2010a). Design of observer-based feedback control for time-delay systems with application to automotive powertrain control, *Journal of Franklin Institute* **347**, 1, pp. 358–376.

Yi, S., Nelson, P. W. and Ulsoy, A. G. (2010b). Eigenvalue assignment via the Lambert W function for control for time-delay systems, *Journal of Vibration and Control* (published online, doi:10.1177/1077546309341102).

Yi, S., Nelson, P. W. and Ulsoy, A. G. (2010c). Robust control and time-domain specifications for systems for delay differential equations via eigenvalue assignment, *Journal of Dynamic Systems, Measurement and Control* **132**, 3, (7 pages, doi:10.1115/1.4001339).

Yi, S. and Ulsoy, A. G. (2006). Solution of a system of linear delay differential equations using the matrix Lambert function, in *Proceedings of 2006 American Control Conference, Minneapolis, MN, 2006 June*, pp. 2433–2438.

Yi, S., Ulsoy, A. G. and Nelson, P. W. (2006b). Solution of systems of linear delay differential equations via Laplace transformation, in *Proceedings of 45th IEEE Conference on Decision and Control, San Diego, CA, DEC. 2006*, pp. 2535–2540.

Yu, X. G. (2008). An LMI approach to robust H_∞ filtering for uncertain systems with time-varying distributed delays, *Journal of the Franklin Institute* **345**, pp. 877–890.

Zafer, N. (2007). Discussion: "Analysis of a system of linear delay differential equations", *Journal of Dynamic Systems Measurement and Control* **129**, 1, pp. 121–122.

Zhong, Q. C. (2006). *Robust Control of Time-Delay Systems* (Springer, London).

Index